Nature-Inspired Metaheuristic Algorithms

Nature-Inspired Metaheuristic Algorithms Second Edition

Xin-She Yang

University of Cambridge, United Kingdom

LUNIVER PRESS

Published in 2010 by Luniver Press
Frome, BA11 6TT, United Kingdom
www.luniver.com

British Library Cataloguing-in-Publication Data
A catalogue record for this book is available from
the British Library

ISBN-13: 978-1-905986-28-6
ISBN-10: 1-905986-28-9

CONTENTS

Preface to the Second Edition

Since the publication of the first edition of this book in 2008, significant developments have been made in metaheuristics, and new nature-inspired metaheuristic algorithms emerge, including cuckoo search and bat algorithms. Many readers have taken time to write to me personally, providing valuable feedback, asking for more details of algorithm implementation, or simply expressing interests in applying these new algorithms in their applications.

In this revised edition, we strive to review the latest developments in metaheuristic algorithms, to incorporate readers' suggestions, and to provide a more detailed description to algorithms. Firstly, we have added detailed descriptions of how to incorporate constraints in the actual implementation. Secondly, we have added three chapters on differential evolution, cuckoo search and bat algorithms, while some existing chapters such as ant algorithms and bee algorithms are combined into one due to their similarity. Thirdly, we also explained artificial neural networks and support vector machines in the framework of optimization and metaheuristics. Finally, we have been trying in this book to provide a consistent and unified approach to metaheuristic algorithms, from a brief history in the first chapter to the unified approach in the last chapter.

Furthermore, we have provided more Matlab programs. At the same time, we also omit some of the implementation such as genetic algorithms, as we know that there are many good software packages (both commercial and open course). This allows us to focus more on the implementation of new algorithms. Some of the programs also have a version for constrained optimization, and readers can modify them for their own applications.

Even with the good intention to cover most popular metaheuristic algorithms, the choice of algorithms is a difficult task, as we do not have the space to cover every algorithm. The omission of an algorithm does not mean that it is not popular. In fact, some algorithms are very powerful and routinely used in many applications. Good examples are Tabu search and combinatorial algorithms, and interested readers can refer to the references provided at the end of the book. The effort in writing this little book becomes worth while if this book could in some way encourage readers' interests in metaheuristics.

Xin-She Yang

August 2010

Preface to the First Edition

Modern metaheuristic algorithms such as the ant colony optimization and the harmony search start to demonstrate their power in dealing with tough optimization problems and even NP-hard problems. This book reviews and introduces the state-of-the-art nature-inspired metaheuristic algorithms in optimization, including genetic algorithms (GA), particle swarm optimization (PSO), simulated annealing (SA), ant colony optimization (ACO), bee algorithms (BA), harmony search (HS), firefly algorithms (FA), photosynthetic algorithm (PA), enzyme algorithm (EA) and Tabu search. By implementing these algorithms in Matlab/Octave, we will use worked examples to show how each algorithm works. This book is thus an ideal textbook for an undergraduate and/or graduate course. As some of the algorithms such as the harmony search and firefly algorithms are at the forefront of current research, this book can also serve as a reference book for researchers.

I would like to thank my editor, Andy Adamatzky, at Luniver Press for his help and professionalism. Last but not least, I thank my wife and son for their help.

Xin-She Yang

Cambridge, 2008

Chapter 1

INTRODUCTION

It is no exaggeration to say that optimization is everywhere, from engineering design to business planning and from the routing of the Internet to holiday planning. In almost all these activities, we are trying to achieve certain objectives or to optimize something such as profit, quality and time. As resources, time and money are always limited in real-world applications, we have to find solutions to optimally use these valuable resources under various constraints. Mathematical optimization or programming is the study of such planning and design problems using mathematical tools. Nowadays, computer simulations become an indispensable tool for solving such optimization problems with various efficient search algorithms.

1.1 OPTIMIZATION

Mathematically speaking, it is possible to write most optimization problems in the generic form

$$\underset{\boldsymbol{x}\in\Re^n}{\text{minimize}} \quad f_i(\boldsymbol{x}), \qquad (i = 1, 2, ..., M), \tag{1.1}$$

$$\text{subject to } h_j(\boldsymbol{x}) = 0, \quad (j = 1, 2, ..., J), \tag{1.2}$$

$$g_k(\boldsymbol{x}) \leq 0, \quad (k = 1, 2, ..., K), \tag{1.3}$$

where $f_i(\boldsymbol{x}), h_j(\boldsymbol{x})$ and $g_k(\boldsymbol{x})$ are functions of the design vector

$$\boldsymbol{x} = (x_1, x_2, ..., x_n)^T. \tag{1.4}$$

Here the components x_i of $\boldsymbol{x}$ are called design or decision variables, and they can be real continuous, discrete or the mixed of these two.

The functions $f_i(\boldsymbol{x})$ where $i = 1, 2, ..., M$ are called the objective functions or simply cost functions, and in the case of $M = 1$, there is only a single objective. The space spanned by the decision variables is called the design space or search space $\Re^n$, while the space formed by the objective function values is called the solution space or response space. The equalities for h_j and inequalities for g_k are called constraints. It is worth pointing out that we can also write the inequalities in the other way ≥ 0, and we can also formulate the objectives as a maximization problem.

In a rare but extreme case where there is no objective at all, there are only constraints. Such a problem is called a feasibility problem because any feasible solution is an optimal solution.

If we try to classify optimization problems according to the number of objectives, then there are two categories: single objective $M = 1$ and multiobjective $M > 1$. Multiobjective optimization is also referred to as multicriteria or even multi-attributes optimization in the literature. In real-world problems, most optimization tasks are multiobjective. Though the algorithms we will discuss in this book are equally applicable to multiobjective optimization with some modifications, we will mainly place the emphasis on single objective optimization problems.

Similarly, we can also classify optimization in terms of number of constraints $J + K$. If there is no constraint at all $J = K = 0$, then it is called an unconstrained optimization problem. If $K = 0$ and $J \geq 1$, it is called an equality-constrained problem, while $J = 0$ and $K \geq 1$ becomes an inequality-constrained problem. It is worth pointing out that in some formulations in the optimization literature, equalities are not explicitly included, and only inequalities are included. This is because an equality can be written as two inequalities. For example $h(\boldsymbol{x}) = 0$ is equivalent to $h(\boldsymbol{x}) \leq 0$ and $h(\boldsymbol{x}) \geq 0$.

We can also use the actual function forms for classification. The objective functions can be either linear or nonlinear. If the constraints h_j and g_k are all linear, then it becomes a linearly constrained problem. If both the constraints and the objective functions are all linear, it becomes a linear programming problem. Here 'programming' has nothing to do with computing programming, it means planning and/or optimization. However, generally speaking, all f_i, h_j and g_k are nonlinear, we have to deal with a nonlinear optimization problem.

1.2 SEARCH FOR OPTIMALITY

After an optimization problem is formulated correctly, the main task is to find the optimal solutions by some solution procedure using the right mathematical techniques.

Figuratively speaking, searching for the optimal solution is like treasure hunting. Imagine we are trying to hunt for a hidden treasure in a hilly landscape within a time limit. In one extreme, suppose we are blindfold without any guidance, the search process is essentially a pure random search, which is usually not efficient as we can expect. In another extreme, if we are told the treasure is placed at the highest peak of a known region, we will then directly climb up to the steepest cliff and try to reach to the highest peak, and this scenario corresponds to the classical hill-climbing techniques. In most cases, our search is between these extremes. We are not blind-fold, and we do not know where to look for. It is a silly idea to search every single square inch of an extremely large hilly region so as to find the treasure.

The most likely scenario is that we will do a random walk, while looking for some hints; we look at some place almost randomly, then move to another plausible place, then another and so on. Such random walk is a main characteristic of modern search algorithms. Obviously, we can either do the treasure-hunting alone, so the whole path is a trajectory-based search, and simulated annealing is such a kind. Alternatively, we can ask a group of people to do the hunting and share the information (and any treasure found), and this scenario uses the so-called swarm intelligence and corresponds to the particle swarm optimization, as we will discuss later in detail. If the treasure is really important and if the area is extremely large, the search process will take a very long time. If there is no time limit and if any region is accessible (for example, no islands in a lake), it is theoretically possible to find the ultimate treasure (the global optimal solution).

Obviously, we can refine our search strategy a little bit further. Some hunters are better than others. We can only keep the better hunters and recruit new ones, this is something similar to the genetic algorithms or evolutionary algorithms where the search agents are improving. In fact, as we will see in almost all modern metaheuristic algorithms, we try to use the best solutions or agents, and randomize (or replace) the not-so-good ones, while evaluating each individual's competence (fitness) in combination with the system history (use of memory). With such a balance, we intend to design better and efficient optimization algorithms.

Classification of optimization algorithm can be carried out in many ways. A simple way is to look at the nature of the algorithm, and this divides the algorithms into two categories: deterministic algorithms, and stochastic algorithms. Deterministic algorithms follow a rigorous procedure, and its path and values of both design variables and the functions are repeatable. For example, hill-climbing is a deterministic algorithm, and for the same starting point, they will follow the same path whether you run the program today or tomorrow. On the other hand, stochastic algorithms always have some randomness. Genetic algorithms are a good example, the strings or solutions in the population will be different each time you run a program since the algorithms use some pseudo-random numbers, though the final results may be no big difference, but the paths of each individual are not exactly repeatable.

Furthermore, there is a third type of algorithm which is a mixture, or a hybrid, of deterministic and stochastic algorithms. For example, hill-climbing with a random restart is a good example. The basic idea is to use the deterministic algorithm, but start with different initial points. This has certain advantages over a simple hill-climbing technique, which may be stuck in a local peak. However, since there is a random component in this hybrid algorithm, we often classify it as a type of stochastic algorithm in the optimization literature.

1.3 NATURE-INSPIRED METAHEURISTICS

Most conventional or classic algorithms are deterministic. For example, the simplex method in linear programming is deterministic. Some deterministic optimization algorithms used the gradient information, they are called gradient-based algorithms. For example, the well-known Newton-Raphson algorithm is gradient-based, as it uses the function values and their derivatives, and it works extremely well for smooth unimodal problems. However, if there is some discontinuity in the objective function, it does not work well. In this case, a non-gradient algorithm is preferred. Non-gradient-based or gradient-free algorithms do not use any derivative, but only the function values. Hooke-Jeeves pattern search and Nelder-Mead downhill simplex are examples of gradient-free algorithms.

For stochastic algorithms, in general we have two types: heuristic and metaheuristic, though their difference is small. Loosely speaking, *heuristic* means 'to find' or 'to discover by trial and error'. Quality solutions to a tough optimization problem can be found in a reasonable amount of time, but there is no guarantee that optimal solutions are reached. It hopes that these algorithms work most of the time, but not all the time. This is good when we do not necessarily want the best solutions but rather good solutions which are easily reachable.

Further development over the heuristic algorithms is the so-called metaheuristic algorithms. Here *meta-* means 'beyond' or 'higher level', and they generally perform better than simple heuristics. In addition, all metaheuristic algorithms use certain tradeoff of randomization and local search. It is worth pointing out that no agreed definitions of heuristics and metaheuristics exist in the literature; some use 'heuristics' and 'metaheuristics' interchangeably. However, the recent trend tends to name all stochastic algorithms with randomization and local search as metaheuristic. Here we will also use this convention. Randomization provides a good way to move away from local search to the search on the global scale. Therefore, almost all metaheuristic algorithms intend to be suitable for global optimization.

Heuristics is a way by trial and error to produce acceptable solutions to a complex problem in a reasonably practical time. The complexity of the problem of interest makes it impossible to search every possible solution or combination, the aim is to find good feasible solution in an acceptable timescale. There is no guarantee that the best solutions can be found, and we even do not know whether an algorithm will work and why if it does work. The idea is to have an efficient but practical algorithm that will work most the time and is able to produce good quality solutions. Among the found quality solutions, it is expected some of them are nearly optimal, though there is no guarantee for such optimality.

Two major components of any metaheuristic algorithms are: intensification and diversification, or exploitation and exploration. Diversification means to generate diverse solutions so as to explore the search space on the global scale, while intensification means to focus on the search in a local region by exploiting the information that a current good solution is found in this region. This is in combination with with the selection of the best solutions. The selection of the best ensures that the solutions will converge to the optimality, while the diversification via randomization avoids the solutions being trapped at local optima and, at the same time, increases the diversity of the solutions. The good combination of these two major components will usually ensure that the global optimality is achievable.

Metaheuristic algorithms can be classified in many ways. One way is to classify them as: population-based and trajectory-based. For example, genetic algorithms are population-based as they use a set of strings, so is the particle swarm optimization (PSO) which uses multiple agents or particles.

On the other hand, simulated annealing uses a single agent or solution which moves through the design space or search space in a piecewise style. A better move or solution is always accepted, while a not-so-good move can be accepted with a certain probability. The steps or moves trace a trajectory in the search space, with a non-zero probability that this trajectory can reach the global optimum.

Before we introduce all popular meteheuristic algorithms in detail, let us look at their history briefly.

1.4 A BRIEF HISTORY OF METAHEURISTICS

Throughout history, especially at the early periods of human history, we humans' approach to problem-solving has always been heuristic or metaheuristic – by trial and error. Many important discoveries were done by 'thinking outside the box', and often by accident; that is heuristics. Archimedes's Eureka moment was a heuristic triumph. In fact, our daily learning experience (at least as a child) is dominantly heuristic.

Despite its ubiquitous nature, metaheuristics as a scientific method to problem solving is indeed a modern phenomenon, though it is difficult to pinpoint when the metaheuristic method was first used. Alan Turing was probably the first to use heuristic algorithms during the second World War when he was breaking German Enigma ciphers at Bletchley Park. Turing called his search method *heuristic search*, as it could be expected it worked most of time, but there was no guarantee to find the correct solution, but it was a tremendous success. In 1945, Turing was recruited to the National Physical Laboratory (NPL), UK where he set out his design for the Automatic Computing Engine (ACE). In an NPL report on *Intelligent machinery* in 1948, he outlined his innovative ideas of machine intelligence and learning, neural networks and evolutionary algorithms.

The 1960s and 1970s were the two important decades for the development of evolutionary algorithms. First, John Holland and his collaborators at the University of Michigan developed the genetic algorithms in 1960s and 1970s. As early as 1962, Holland studied the adaptive system and was the first to use crossover and recombination manipulations for modeling such system. His seminal book summarizing the development of genetic algorithms was published in 1975. In the same year, De Jong finished his important dissertation showing the potential and power of genetic algorithms for a wide range of objective functions, either noisy, multimodal or even discontinuous.

In essence, a genetic algorithm (GA) is a search method based on the abstraction of Darwinian evolution and natural selection of biological systems and representing them in the mathematical operators: crossover or recombination, mutation, fitness, and selection of the fittest. Ever since, genetic algorithms become so successful in solving a wide range of optimization problems, there have several thousands of research articles and hundreds of books written. Some statistics show that a vast majority of Fortune 500 companies are now using them routinely to solve tough combinatorial optimization problems such as planning, data-fitting, and scheduling.

During the same period, Ingo Rechenberg and Hans-Paul Schwefel both then at the Technical University of Berlin developed a search technique for solving optimization problem in aerospace engineering, called evolutionary strategy, in 1963. Later, Peter Bienert joined them and began to construct an automatic experimenter using simple rules of mutation and selection. There was no crossover in this technique, only mutation was used to produce an offspring and an improved solution was kept at each generation. This was essentially a simple trajectory-style hill-climbing algorithm with randomization. As early as 1960, Lawrence J. Fogel intended to use simulated evolution as a learning process as a tool to study artificial intelligence. Then, in 1966, L. J. Fogel, together A. J. Owen and M. J. Walsh, developed the evolutionary programming technique by representing solutions as finite-state machines and randomly mutating one of these machines. The above innovative ideas and methods have evolved into a much wider discipline, called *evolutionary algorithms* and/or *evolutionary computation*.

Although our focus in this book is metaheuristic algorithms, other algorithms can be thought as a heuristic optimization technique. These includes artificial neural networks, support vector machines and many other machine learning techniques. Indeed, they all intend to minimize their learning errors and prediction (capability) errors via iterative trials and errors.

Artificial neural networks are now routinely used in many applications. In 1943, W. McCulloch and W. Pitts proposed the artificial neurons as simple information processing units. The concept of a neural network was probably first proposed by Alan Turing in his 1948 NPL report concerning 'intelligent machinery'. Significant developments were carried out from the 1940s and 1950s to the 1990s with more than 60 years of history.

The support vector machine as a classification technique can date back to the earlier work by V. Vapnik in 1963 on linear classifiers, and the nonlinear classification with kernel techniques were developed by V. Vapnik and his collaborators in the 1990s. A systematical summary in Vapnik's book on the Nature of Statistical Learning Theory was published in 1995.

The two decades of 1980s and 1990s were the most exciting time for metaheuristic algorithms. The next big step is the development of simulated annealing (SA) in 1983, an optimization technique, pioneered by S. Kirkpatrick, C. D. Gellat and M. P. Vecchi, inspired by the annealing process of metals. It is a trajectory-based search algorithm starting with an initial guess solution at a high temperature, and gradually cooling down the system. A move or new solution is accepted if it is better; otherwise, it is accepted with a probability, which makes it possible for the system to escape any local optima. It is then expected that if the system is cooled down slowly enough, the global optimal solution can be reached.

The actual first usage of memory in modern metaheuristics is probably due to Fred Glover's Tabu search in 1986, though his seminal book on Tabu search was published later in 1997.

In 1992, Marco Dorigo finished his PhD thesis on optimization and natural algorithms, in which he described his innovative work on ant colony optimization (ACO). This search technique was inspired by the swarm intelligence of social ants using pheromone as a chemical messenger. Then, in 1992, John R. Koza of Stanford University published a treatise on genetic programming which laid the foundation of a whole new area of machine learning, revolutionizing computer programming. As early as in 1988, Koza applied his first patent on genetic programming. The basic idea is to use the genetic principle to breed computer programs so as to gradually produce the best programs for a given type of problem.

Slightly later in 1995, another significant progress is the development of the particle swarm optimization (PSO) by American social psychologist James Kennedy, and engineer Russell C. Eberhart. Loosely speaking, PSO is an optimization algorithm inspired by swarm intelligence of fish and birds and by even human behavior. The multiple agents, called particles, swarm around the search space starting from some initial random guess. The swarm communicates the current best and shares the global best so as to focus on the quality solutions. Since its development, there have been about 20 different variants of particle swarm optimization techniques, and have been applied to almost all areas of tough optimization problems. There is some strong evidence that PSO is better than traditional search algorithms and even better than genetic algorithms for many types of problems, though this is far from conclusive.

In around 1996 and later in 1997, R. Storn and K. Price developed their vector-based evolutionary algorithm, called differential evolution (DE), and this algorithm proves more efficient than genetic algorithms in many applications.

In 1997, the publication of the 'no free lunch theorems for optimization' by D. H. Wolpert and W. G. Macready sent out a shock way to the optimization community. Researchers have been always trying to find better algorithms, or even universally robust algorithms, for optimization, especially for tough NP-hard optimization problems. However, these theorems state that if algorithm A performs better than algorithm B for some optimization functions, then B will outperform A for other functions. That is to say, if averaged over all possible function space, both algorithms A and B will perform on average equally well. Alternatively, there is no universally better algorithms exist. That is disappointing, right? Then, people realized that we do not need the average over all possible functions for a given optimization problem. What we want is to find the best solutions, which has nothing to do with average over all possible function space. In addition, we can accept the fact that there is no universal or magical tool, but we do know from our experience that some algorithms indeed outperform others for given types of optimization problems. So the research now focuses on finding the best and most efficient algorithm(s) for a given problem. The objective is to design better algorithms for most types of problems, not for all the problems. Therefore, the search is still on.

At the turn of the 21st century, things became even more exciting. First, Zong Woo Geem *et al.* in 2001 developed the harmony search (HS) algorithm, which has been widely applied in solving various optimization problems such as water distribution, transport modelling and scheduling. In 2004, S. Nakrani and C. Tovey proposed the honey bee algorithm and its application for optimizing Internet hosting centers, which followed by the development of a novel bee algorithm by D. T. Pham *et al.* in 2005 and the artificial bee colony (ABC) by D. Karaboga in 2005. In 2008, the author of this book developed the firefly algorithm (FA)[1]. Quite a few research articles on the firefly algorithm then followed, and this algorithm has attracted a wide range of interests. In 2009, Xin-She Yang at Cambridge University, UK, and Suash Deb at Raman College of Engineering, India, introduced an efficient cuckoo search (CS) algorithm, and it has been demonstrated that CS is far more effective than most existing metaheuristic algorithms

[1]X. S. Yang, *Nature-Inspired Meteheuristic Algorithms*, Luniver Press, (2008)

including particle swarm optimization[2]. In 2010, the author of this book developed a bat-inspired algorithm for continuous optimization, and its efficiency is quite promising.

As we can see, more and more metaheuristic algorithms are being developed. Such a diverse range of algorithms necessitates a systematic summary of various metaheuristic algorithms, and this book is such an attempt to introduce all the latest nature-inspired metaheuristics with diverse applications.

We will discuss all major modern metaheuristic algorithms in the rest of this book, including simulated annealing (SA), genetic algorithms (GA), ant colony optimization (ACO), bee algorithms (BA), differential evolution (DE), particle swarm optimization (PSO), harmony search (HS), the firefly algorithm (FA), cuckoo search (CS) and bat-inspired algorithm (BA), and others.

REFERENCES

1. C. M. Bishop, *Neural Networks for Pattern Recognition*, Oxford University Press, Oxford, 1995.
2. B. J. Copeland, *The Essential Turing*, Oxford University Press, 2004.
3. B. J. Copeland, *Alan Turing's Automatic Computing Engine*, Oxford University Press, 2005.
4. K. De Jong, *Analysis of the Behaviour of a Class of Genetic Adaptive Systems*, PhD thesis, University of Michigan, Ann Anbor, 1975.
5. M. Dorigo, *Optimization, Learning and Natural Algorithms*, PhD thesis, Politecnico di Milano, Italy, 1992.
6. L. J. Fogel, A. J. Owens, and M. J. Walsh, *Artificial Intelligence Through Simulated Evolution*, Wiley, 1966.
7. Z. W. Geem, J. H. Kim and G. V. Loganathan, A new heuristic optimization: Harmony search, *Simulation*, **76**(2), 60-68 (2001).
8. F. Glover and M. Laguna, *Tabu Search*, Kluwer Academic Publishers, Boston, 1997.
9. J. Holland, *Adaptation in Natural and Artificial systems*, University of Michigan Press, Ann Anbor, 1975.
10. P. Judea, *Heuristics*, Addison-Wesley, 1984.
11. D. Karaboga, An idea based on honey bee swarm for numerical optimization, Technical Report, Erciyes University, 2005.
12. J. Kennedy and R. Eberhart, Particle swarm optimization, in: *Proc. of the IEEE Int. Conf. on Neural Networks*, Piscataway, NJ, pp. 1942-1948 (1995).

[2]Novel cuckoo search 'beats' particle swarm optimization, *Science Daily*, news article (28 May 2010), www.sciencedaily.com

13. S. Kirkpatrick, C. D. Gellat, and M. P. Vecchi, Optimization by simulated annealing, *Science*, **220**, 671-680 (1983).

14. J. R. Koza, *Genetic Programming: One the Programming of Computers by Means of Natural Selection*, MIT Press, 1992.

15. S. Nakrani and C. Tovey, On honey bees and dynamic server allocation in Internet hostubg centers, *Adaptive Behavior*, **12**, 223-240 (2004).

16. D. T. Pham, A. Ghanbarzadeh, E. Koc, S. Otri, S. Rahim and M. Zaidi, The bees algorithm, Technical Note, Manufacturing Engineering Center, Cardiff University, 2005.

17. A. Schrijver, On the history of combinatorial optimization (till 1960), in: *Handbook of Discrete Optimization* (Eds K. Aardal, G. L. Nemhauser, R. Weismantel), Elsevier, Amsterdam, pp.1-68 (2005).

18. H. T. Siegelmann and E. D. Sontag, Turing computability with neural nets, *Appl. Math. Lett.*, **4**, 77-80 (1991).

19. R. Storn and K. Price, Differential evolution - a simple and efficient heuristic for global optimization over continuous spaces, *Journal of Global Optimization*, **11**, 341-359 (1997).

20. A. M. Turing, *Intelligent Machinery*, National Physical Laboratory, technical report, 1948.

21. V. Vapnik, *The Nature of Statistical Learning Theory*, Springer-Verlag, New York, 1995.

22. V. Vapnik, S. Golowich, A. Smola, Support vector method for function approxiation, regression estimation, and signal processing, in: *Advances in Neural Information Processing System 9* (Eds. M. Mozer, M. Jordan and T. Petsche), MIT Press, Cambridge MA, 1997.

23. D. H. Wolpert and W. G. Macready, No free lunch theorems for optimization, *IEEE Transaction on Evolutionary Computation*, **1**, 67-82 (1997).

24. X. S. Yang, *Nature-Inspired Metaheuristic Algorithms*, Luniver Press, (2008).

25. X. S. Yang, Firefly algorithms for multimodal optimization, *Proc. 5th Symposium on Stochastic Algorithms, Foundations and Applications, SAGA 2009*, Eds. O. Watanabe and T. Zeugmann, Lecture Notes in Computer Science, **5792**, 169-178 (2009).

26. X. S. Yang and S. Deb, Cuckoo search via Lévy flights, in: *Proc. of World Congress on Nature & Biologically Inspired Computing* (NaBic 2009), IEEE Publications, USA, pp. 210-214 (2009).

27. X. S. Yang and S. Deb, Engineering optimization by cuckoo search, *Int. J. Math. Modelling & Num. Optimization*, **1**, 330-343 (2010).

28. X. S. Yang, A new metaheuristic bat-inspired algorithm, in: *Nature Inspired Cooperative Strategies for Optimization* (NICSO 2010) (Eds. J. R. Gonzalez et al.), Springer, SCI **284**, 65-74 (2010).

29. History of optimization, http://hse-econ.fi/kitti/opthist.html

30. Turing Archive for the History of Computing, www.alanturing.net/

Chapter 2

RANDOM WALKS AND LÉVY FLIGHTS

From the brief analysis of the main characteristics of metaheuristic algorithms in the first chapter, we know that randomization plays an important role in both exploration and exploitation, or diversification and intensification. The essence of such randomization is the random walk. In this chapter, we will briefly review the fundamentals of random walks, Lévy flights and Markov chains. These concepts may provide some hints and insights into how and why metaheuristic algorithms behave.

2.1 RANDOM VARIABLES

Loosely speaking, a random variable can be considered as an expression whose value is the realization or outcome of events associated with a random process such as the noise level on the street. The values of random variables are real, though for some variables such as the number of cars on a road can only take discrete values, and such random variables are called discrete random variables. If a random variable such as noise at a particular location can take any real values in an interval, it is called continuous. If a random variable can have both continuous and discrete values, it is called a mixed type. Mathematically speaking, a random variable is a function which maps events to real numbers. The domain of this mapping is called the sample space.

For each random variable, a probability density function can be used to express its probability distribution. For example, the number of phone calls per minute, and the number of users of a web server per day all obey the Poisson distribution

$$p(n;\lambda) = \frac{\lambda^n e^{-\lambda}}{n!}, \qquad (n = 0, 1, 2, ...), \tag{2.1}$$

where $\lambda > 0$ is a parameter which is the mean or expectation of the occurrence of the event during a unit interval.

Different random variables will have different distributions. Gaussian distribution or normal distribution is by far the most popular distributions, because many physical variables including light intensity, and errors/uncertainty in measurements, and many other processes obey the normal distribution

$$p(x;\mu,\sigma^2) = \frac{1}{\sigma\sqrt{2\pi}} \exp[-\frac{(x-\mu)^2}{2\sigma^2}], \qquad -\infty < x < \infty, \tag{2.2}$$

where μ is the mean and $\sigma > 0$ is the standard deviation. This normal distribution is often denoted by $\mathrm{N}(\mu, \sigma^2)$. In the special case when $\mu = 0$ and $\sigma = 1$, it is called a standard normal distribution, denoted by $\mathrm{N}(0, 1)$.

In the context of metaheuristics, another important distribution is the so-called Lévy distribution, which is a distribution of the sum of N identically and independently distribution random variables whose Fourier transform takes the following form

$$F_N(k) = \exp[-N|k|^\beta]. \tag{2.3}$$

The inverse to get the actual distribution $L(s)$ is not straightforward, as the integral

$$L(s) = \frac{1}{\pi} \int_0^\infty \cos(\tau s) e^{-\alpha\, \tau^\beta} d\tau, \qquad (0 < \beta \leq 2), \tag{2.4}$$

does not have analytical forms, except for a few special cases. Here $L(s)$ is called the Lévy distribution with an index β. For most applications, we can set $\alpha = 1$ for simplicity. Two special cases are $\beta = 1$ and $\beta = 2$. When $\beta = 1$, the above integral becomes the Cauchy distribution. When $\beta = 2$, it becomes the normal distribution. In this case, Lévy flights become the standard Brownian motion.

Mathematically speaking, we can express the integral (2.4) as an asymptotic series, and its leading-order approximation for the flight length results in a power-law distribution

$$L(s) \sim |s|^{-1-\beta}, \tag{2.5}$$

which is heavy-tailed. The variance of such a power-law distribution is infinite for $0 < \beta < 2$. The moments diverge (or are infinite) for $0 < \beta < 2$, which is a stumbling block for mathematical analysis.

2.2 RANDOM WALKS

A random walk is a random process which consists of taking a series of consecutive random steps. Mathematically speaking, let S_N denotes the sum of each consecutive random step X_i, then S_N forms a random walk

$$S_N = \sum_{i=1}^{N} X_i = X_1 + ... + X_N, \tag{2.6}$$

where X_i is a random step drawn from a random distribution. This relationship can also be written as a recursive formula

$$S_N = \sum_{i=1}^{N-1} + X_N = S_{N-1} + X_N, \tag{2.7}$$

which means the next state S_N will only depend the current existing state S_{N-1} and the motion or transition X_N from the existing state to the next state. This is typically the main property of a Markov chain to be introduced later.

Here the step size or length in a random walk can be fixed or varying. Random walks have many applications in physics, economics, statistics, computer sciences, environmental science and engineering.

Consider a scenario, a drunkard walks on a street, at each step, he can randomly go forward or backward, this forms a random walk in one-dimensional. If this drunkard walks on a football pitch, he can walk in any direction randomly, this becomes a 2D random walk. Mathematically speaking, a random walk is given by the following equation

$$S_{t+1} = S_t + w_t, \tag{2.8}$$

where S_t is the current location or state at t, and w_t is a step or random variable with a known distribution.

If each step or jump is carried out in the n-dimensional space, the random walk discussed earlier

$$S_N = \sum_{i=1}^{N} X_i, \tag{2.9}$$

becomes a random walk in higher dimensions. In addition, there is no reason why each step length should be fixed. In fact, the step size can also vary according to a known distribution. If the step length obeys the Gaussian distribution, the random walk becomes the Brownian motion (see Fig. 2.1).

In theory, as the number of steps N increases, the central limit theorem implies that the random walk (2.9) should approaches a Gaussian distribution. As the mean of particle locations shown in Fig. 2.1 is obviously zero, their variance will increase linearly with t. In general, in the d-dimensional space, the variance of Brownian random walks can be written as

$$\sigma^2(t) = |v_0|^2 t^2 + (2dD)t, \tag{2.10}$$

where v_0 is the drift velocity of the system. Here $D = s^2/(2\tau)$ is the effective diffusion coefficient which is related to the step length s over a short time interval τ during each jump.

Figure 2.1: Brownian motion in 2D: random walk with a Gaussian step-size distribution and the path of 50 steps starting at the origin $(0, 0)$ (marked with $\bullet$).

Therefore, the Brownian motion $B(t)$ essentially obeys a Gaussian distribution with zero mean and time-dependent variance. That is, $B(t) \sim N(0, \sigma^2(t))$ where $\sim$ means the random variable obeys the distribution on the right-hand side; that is, samples should be drawn from the distribution. A diffusion process can be viewed as a series of Brownian motion, and the motion obeys the Gaussian distribution. For this reason, standard diffusion is often referred to as the Gaussian diffusion. If the motion at each step is not Gaussian, then the diffusion is called non-Gaussian diffusion.

If the step length obeys other distribution, we have to deal with more generalized random walk. A very special case is when the step length obeys the Lévy distribution, such a random walk is called a Lévy flight or Lévy walk.

2.3 LÉVY DISTRIBUTION AND LÉVY FLIGHTS

Broadly speaking, Lévy flights are a random walk whose step length is drawn from the Lévy distribution, often in terms of a simple power-law formula $L(s) \sim |s|^{-1-\beta}$ where $0 < \beta \leq 2$ is an index. Mathematically speaking, a simple version of Lévy distribution can be defined as

$$L(s, \gamma, \mu) = \begin{cases} \sqrt{\frac{\gamma}{2\pi}} \exp[-\frac{\gamma}{2(s-\mu)}] \frac{1}{(s-\mu)^{3/2}}, & 0 < \mu < s < \infty \\ 0 & \text{otherwise}, \end{cases} \tag{2.11}$$

where $\mu > 0$ is a minimum step and γ is a scale parameter. Clearly, as $s \to \infty$, we have

$$L(s, \gamma, \mu) \approx \sqrt{\frac{\gamma}{2\pi}} \frac{1}{s^{3/2}}. \tag{2.12}$$

This is a special case of the generalized Lévy distribution.

Figure 2.2: Lévy flights in consecutive 50 steps starting at the origin (0, 0) (marked with •).

In general, Lévy distribution should be defined in terms of Fourier transform

$$F(k) = \exp[-\alpha|k|^{\beta}], \qquad 0 < \beta \leq 2, \tag{2.13}$$

where α is a scale parameter. The inverse of this integral is not easy, as it does not have analytical form, except for a few special cases.

For the case of $\beta = 2$, we have

$$F(k) = \exp[-\alpha k^2], \tag{2.14}$$

whose inverse Fourier transform corresponds to a Gaussian distribution. Another special case is $\beta = 1$, and we have

$$F(k) = \exp[-\alpha|k|], \tag{2.15}$$

which corresponds to a Cauchy distribution

$$p(x, \gamma, \mu) = \frac{1}{\pi} \frac{\gamma}{\gamma^2 + (x - \mu)^2}, \tag{2.16}$$

where μ is the location parameter, while γ controls the scale of this distribution.

For the general case, the inverse integral

$$L(s) = \frac{1}{\pi} \int_0^{\infty} \cos(ks) \exp[-\alpha|k|^{\beta}] dk, \tag{2.17}$$

can be estimated only when s is large. We have

$$L(s) \to \frac{\alpha\, \beta\, \Gamma(\beta) \sin(\pi\beta/2)}{\pi|s|^{1+\beta}}, \qquad s \to \infty. \tag{2.18}$$

Here $\Gamma(z)$ is the Gamma function

$$\Gamma(z) = \int_0^\infty t^{z-1} e^{-t} dt. \tag{2.19}$$

In the case when $z = n$ is an integer, we have $\Gamma(n) = (n-1)!$.

Lévy flights are more efficient than Brownian random walks in exploring unknown, large-scale search space. There are many reasons to explain this efficiency, and one of them is due to the fact that the variance of Lévy flights

$$\sigma^2(t) \sim t^{3-\beta}, \qquad 1 \le \beta \le 2, \tag{2.20}$$

increases much faster than the linear relationship (i.e., $\sigma^2(t) \sim t$) of Brownian random walks.

Fig. 2.2 shows the path of Lévy flights of 50 steps starting from $(0, 0)$ with $\beta = 1$. It is worth pointing out that a power-law distribution is often linked to some scale-free characteristics, and Lévy flights can thus show self-similarity and fractal behavior in the flight patterns.

From the implementation point of view, the generation of random numbers with Lévy flights consists of two steps: the choice of a random direction and the generation of steps which obey the chosen Lévy distribution. The generation of a direction should be drawn from a uniform distribution, while the generation of steps is quite tricky. There are a few ways of achieving this, but one of the most efficient and yet straightforward ways is to use the so-called Mantegna algorithm for a symmetric Lévy stable distribution. Here 'symmetric' means that the steps can be positive and negative.

A random variable U and its probability distribution can be called stable if a linear combination of its two identical copies (or U_1 and U_2) obeys the same distribution. That is, $aU_1 + bU_2$ has the same distribution as $cU + d$ where $a, b > 0$ and $c, d \in \Re$. If $d = 0$, it is called strictly stable. Gaussian, Cauchy and Lévy distributions are all stable distributions.

In Mantegna's algorithm, the step length s can be calculated by

$$s = \frac{u}{|v|^{1/\beta}}, \tag{2.21}$$

where u and v are drawn from normal distributions. That is

$$u \sim N(0, \sigma_u^2), \qquad v \sim N(0, \sigma_v^2), \tag{2.22}$$

where

$$\sigma_u = \left\{ \frac{\Gamma(1+\beta) \sin(\pi\beta/2)}{\Gamma[(1+\beta)/2] \, \beta \, 2^{(\beta-1)/2}} \right\}^{1/\beta}, \qquad \sigma_v = 1. \tag{2.23}$$

This distribution (for s) obeys the expected Lévy distribution for $|s| \ge |s_0|$ where s_0 is the smallest step. In principle, $|s_0| \gg 0$, but in reality s_0 can be taken as a sensible value such as $s_0 = 0.1$ to 1.

Studies show that Lévy flights can maximize the efficiency of resource searches in uncertain environments. In fact, Lévy flights have been observed among foraging patterns of albatrosses and fruit flies, and spider monkeys. Even humans such as the Ju/'hoansi hunter-gatherers can trace paths of Lévy-flight patterns. In addition, Lévy flights have many applications. Many physical phenomena such as the diffusion of fluorescent molecules, cooling behavior and noise could show Lévy-flight characteristics under the right conditions.

2.4 OPTIMIZATION AS MARKOV CHAINS

In every aspect, a simple random walk we discussed earlier can be considered as a Markov chain. Briefly speaking, a random variable ζ is a Markov process if the transition probability, from state $\zeta_t = S_i$ at time t to another state $\zeta_{t+1} = S_j$, depends only on the current state ζ_i. That is

$$P(i,j) \equiv P(\zeta_{t+1} = S_j | \zeta_0 = S_p, ..., \zeta_t = S_i)$$
$$= P(\zeta_{t+1} = S_j | \zeta_t = S_i), \tag{2.24}$$

which is independent of the states before t. In addition, the sequence of random variables $(\zeta_0, \zeta_1, ..., \zeta_n)$ generated by a Markov process is subsequently called a Markov chain. The transition probability $P(i,j) \equiv P(i \to j) = P_{ij}$ is also referred to as the transition kernel of the Markov chain.

If we rewrite the random walk relationship (2.7) with a random move governed by w_t which depends on the transition probability P, we have

$$S_{t+1} = S_t + w_t, \tag{2.25}$$

which indeed has the properties of a Markov chain. Therefore, a random walk is a Markov chain.

In order to solve an optimization problem, we can search the solution by performing a random walk starting from a good initial but random guess solution. However, simple or blind random walks are not efficient. To be computationally efficient and effective in searching for new solutions, we have to keep the best solutions found so far, and to increase the mobility of the random walk so as to explore the search space more effectively. Most importantly, we have to find a way to control the walk in such a way that it can move towards the optimal solutions more quickly, rather than wander away from the potential best solutions. These are the challenges for most metaheuristic algorithms.

Further research along the route of Markov chains is that the development of the Markov chain Monte Carlo (MCMC) method, which is a class of sample-generating methods. It attempts to directly draw samples from some highly complex multi-dimensional distribution using a Markov chain with known transition probability. Since the 1990s, the Markov chain Monte Carlo has become a powerful tool for Bayesian statistical analysis, Monte Carlo simulations, and potentially optimization with high nonlinearity.

An important link between MCMC and optimization is that some heuristic and metaheuristic search algorithms such as simulated annealing to be introduced later use a trajectory-based approach. They start with some initial (random) state, and propose a new state (solution) randomly. Then, the move is accepted or not, depending on some probability. There is strongly similar to a Markov chain. In fact, the standard simulated annealing is a random walk.

Mathematically speaking, a great leap in understanding metaheuristic algorithms is to view a Markov chain Monte carlo as an optimization procedure. If we want to find the minimum of an objective function $f(\theta)$ at $\theta = \theta_*$ so that $f_* = f(\theta_*) \leq f(\theta)$, we can convert it to a target distribution for a Markov chain

$$\pi(\theta) = e^{-\beta f(\theta)}, \tag{2.26}$$

where $\beta > 0$ is a parameter which acts as a normalized factor. β value should be chosen so that the probability is close to 1 when $\theta \to \theta_*$. At $\theta = \theta_*$, $\pi(\theta)$ should reach a maximum $\pi_* = \pi(\theta_*) \geq \pi(\theta)$. This requires that the formulation of $L(\theta)$ should be non-negative, which means that some objective functions can be shifted by a large constant $A > 0$ such as $f \leftarrow f + A$ if necessary.

By constructing a Markov chain Monte Carlo, we can formulate a generic framework as outlined by Ghate and Smith in 2008, as shown in Figure 2.3. In this framework, simulated annealing and its many variants are simply a special case with

$$P_t = \begin{cases} \exp[-\frac{\Delta f}{T_t}] & \text{if } f_{t+1} > f_t \\ \\ 1 & \text{if } f_{t+1} \leq f_t \end{cases},$$

In this case, only the difference Δf between the function values is important.

Algorithms such as simulated annealing, to be discussed in the next chapter, use a single Markov chain, which may not be very efficient. In practice, it is usually advantageous to use multiple Markov chains in parallel to increase the overall efficiency. In fact, the algorithms such as particle swarm optimization can be viewed as multiple interacting Markov chains, though such theoretical analysis remains almost intractable. The theory of interacting Markov chains is complicated and yet still under development, however, any progress in such areas will play a central role in the understanding how population- and trajectory-based metaheuristic algorithms perform under various conditions. However, even though we do not fully understand why metaheuristic algorithms work, this does not hinder us to

Markov Chain Algorithm for Optimization

Start with $\zeta_0 \in S$*, at* $t = 0$
while *(criterion)*
Propose a new solution Y_{t+1}*;*
Generate a random number $0 \le P_t \le 1$*;*

$$\zeta_{t+1} = \begin{cases} Y_{t+1} & \textit{with probability } P_t \\ \zeta_t & \textit{with probability } 1 - P_t \end{cases} \tag{2.27}$$

end

Figure 2.3: Optimization as a Markov chain.

use these algorithms efficiently. On the contrary, such mysteries can drive and motivate us to pursue further research and development in metaheuristics.

REFERENCES

1. W. J. Bell, *Searching Behaviour: The Behavioural Ecology of Finding Resources*, Chapman & Hall, London, (1991).
2. C. Blum and A. Roli, Metaheuristics in combinatorial optimization: Overview and conceptual comparison, *ACM Comput. Surv.*, **35**, 268-308 (2003).
3. G. S. Fishman, *Monte Carlo: Concepts, Algorithms and Applications*, Springer, New York, (1995).
4. D. Gamerman, *Markov Chain Monte Carlo*, Chapman & Hall/CRC, (1997).
5. L. Gerencser, S. D. Hill, Z. Vago, and Z. Vincze, Discrete optimization, SPSA, and Markov chain Monte Carlo methods, *Proc. 2004 Am. Contr. Conf.*, 3814-3819 (2004).
6. C. J. Geyer, Practical Markov Chain Monte Carlo, *Statistical Science*, **7**, 473-511 (1992).
7. A. Ghate and R. Smith, Adaptive search with stochastic acceptance probabilities for global optimization, *Operations Research Lett.*, **36**, 285-290 (2008).
8. W. R. Gilks, S. Richardson, and D. J. Spiegelhalter, *Markov Chain Monte Carlo in Practice*, Chapman & Hall/CRC, (1996).
9. M. Gutowski, Lévy flights as an underlying mechanism for global optimization algorithms, *ArXiv Mathematical Physics e-Prints*, June, (2001).
10. W. K. Hastings, Monte Carlo sampling methods using Markov chains and their applications, *Biometrika*, **57**, 97-109 (1970).
11. S. Kirkpatrick, C. D. Gellat and M. P. Vecchi, Optimization by simulated annealing, *Science*, **220**, 670-680 (1983).

12. R. N. Mantegna, Fast, accurate algorithm for numerical simulation of Levy stable stochastic processes, *Physical Review E*, **49**, 4677-4683 (1994).

13. E. Marinari and G. Parisi, Simulated tempering: a new Monte Carlo scheme, *Europhysics Lett.*, **19**, 451-458 (1992).

14. J. P. Nolan, Stable distributions: models for heavy-tailed data, American University, (2009).

15. N. Metropolis, and S. Ulam, The Monte Carlo method, *J. Amer. Stat. Assoc.*, **44**, 335-341 (1949).

16. N. Metropolis, A. W. Rosenbluth, M. N. Rosenbluth, A. H. Teller, and E. Teller, Equation of state calculations by fast computing machines, *J. Chem. Phys.*, **21**, 1087-1092 (1953).

17. I. Pavlyukevich, Lévy flights, non-local search and simulated annealing, *J. Computational Physics*, **226**, 1830-1844 (2007).

18. G. Ramos-Fernandez, J. L. Mateos, O. Miramontes, G. Cocho, H. Larralde, B. Ayala-Orozco, Lévy walk patterns in the foraging movements of spider monkeys (*Ateles geoffroyi*),*Behav. Ecol. Sociobiol.*, **55**, 223-230 (2004).

19. A. M. Reynolds and M. A. Frye, Free-flight odor tracking in Drosophila is consistent with an optimal intermittent scale-free search, *PLoS One*, **2**, e354 (2007).

20. A. M. Reynolds and C. J. Rhodes, The Lévy flight paradigm: random search patterns and mechanisms, *Ecology*, **90**, 877-887 (2009).

21. I. M. Sobol, *A Primer for the Monte Carlo Method*, CRC Press, (1994).

22. M. E. Tipping M. E., Bayesian inference: An introduction to principles and and practice in machine learning, in: *Advanced Lectures on Machine Learning*, O. Bousquet, U. von Luxburg and G. Ratsch (Eds), pp.41-62 (2004).

23. G. M. Viswanathan, S. V. Buldyrev, S. Havlin, M. G. E. da Luz, E. P. Raposo, and H. E. Stanley, Lévy flight search patterns of wandering albatrosses, *Nature*, **381**, 413-415 (1996).

24. E. Weisstein, http://mathworld.wolfram.com

Chapter 3

SIMULATED ANNEALING

One of the earliest and yet most popular metaheuristic algorithms is simulated annealing (SA), which is a trajectory-based, random search technique for global optimization. It mimics the annealing process in material processing when a metal cools and freezes into a crystalline state with the minimum energy and larger crystal size so as to reduce the defects in metallic structures. The annealing process involves the careful control of temperature and its cooling rate, often called annealing schedule.

3.1 ANNEALING AND BOLTZMANN DISTRIBUTION

Since the first development of simulated annealing by Kirkpatrick, Gelatt and Vecchi in 1983, SA has been applied in almost every area of optimization. Unlike the gradient-based methods and other deterministic search methods which have the disadvantage of being trapped into local minima, the main advantage of simulated annealing is its ability to avoid being trapped in local minima. In fact, it has been proved that simulated annealing will converge to its global optimality if enough randomness is used in combination with very slow cooling. Essentially, simulated annealing is a search algorithm via a Markov chain, which converges under appropriate conditions.

Metaphorically speaking, this is equivalent to dropping some bouncing balls over a landscape, and as the balls bounce and lose energy, they settle down to some local minima. If the balls are allowed to bounce enough times and lose energy slowly enough, some of the balls will eventually fall into the globally lowest locations, hence the global minimum will be reached.

The basic idea of the simulated annealing algorithm is to use random search in terms of a Markov chain, which not only accepts changes that improve the objective function, but also keeps some changes that are not ideal. In a minimization problem, for example, any better moves or changes that decrease the value of the objective function f will be accepted; however, some changes that increase f will also be accepted with a probability p. This probability p, also called the transition probability, is determined by

$$p = e^{-\frac{\Delta E}{k_B T}}, \tag{3.1}$$

where k_B is the Boltzmann's constant, and for simplicity, we can use k to denote k_B because $k = 1$ is often used. T is the temperature for controlling the annealing process. ΔE is the change of the energy level. This transition probability is based on the Boltzmann distribution in statistical mechanics.

The simplest way to link ΔE with the change of the objective function Δf is to use

$$\Delta E = \gamma \Delta f, \tag{3.2}$$

where γ is a real constant. For simplicity without losing generality, we can use $k_B = 1$ and $\gamma = 1$. Thus, the probability p simply becomes

$$p(\Delta f, T) = e^{-\Delta f/T}. \tag{3.3}$$

Whether or not we accept a change, we usually use a random number r as a threshold. Thus, if $p > r$, or

$$p = e^{-\frac{\Delta f}{T}} > r, \tag{3.4}$$

the move is accepted.

3.2 PARAMETERS

Here the choice of the right initial temperature is crucially important. For a given change Δf, if T is too high ($T \to \infty$), then $p \to 1$, which means almost all the changes will be accepted. If T is too low ($T \to 0$), then any $\Delta f > 0$ (worse solution) will rarely be accepted as $p \to 0$ and thus the diversity of the solution is limited, but any improvement Δf will almost always be accepted. In fact, the special case $T \to 0$ corresponds to the gradient-based method because only better solutions are accepted, and the system is essentially climbing up or descending along a hill. Therefore, if T is too high, the system is at a high energy state on the topological landscape, and the minima are not easily reached. If T is too low, the system may be trapped in a local minimum (not necessarily the global minimum), and there is not enough energy for the system to jump out the local minimum to explore other minima including the global minimum. So a proper initial temperature should be calculated.

Another important issue is how to control the annealing or cooling process so that the system cools down gradually from a higher temperature to ultimately freeze to a global minimum state. There are many ways of controlling the cooling rate or the decrease of the temperature.

Two commonly used annealing schedules (or cooling schedules) are: linear and geometric. For a linear cooling schedule, we have

$$T = T_0 - \beta t, \tag{3.5}$$

or $T \to T - \delta T$, where T_0 is the initial temperature, and t is the pseudo time for iterations. β is the cooling rate, and it should be chosen in such a way that $T \to 0$ when $t \to t_f$ (or the maximum number N of iterations), this usually gives $\beta = (T_0 - T_f)/t_f$.

On the other hand, a geometric cooling schedule essentially decreases the temperature by a cooling factor $0 < \alpha < 1$ so that T is replaced by αT or

$$T(t) = T_0 \alpha^t, \qquad t = 1, 2, ..., t_f. \tag{3.6}$$

The advantage of the second method is that $T \to 0$ when $t \to \infty$, and thus there is no need to specify the maximum number of iterations. For this reason, we will use this geometric cooling schedule. The cooling process should be slow enough to allow the system to stabilize easily. In practise, $\alpha = 0.7 \sim 0.99$ is commonly used.

In addition, for a given temperature, multiple evaluations of the objective function are needed. If too few evaluations, there is a danger that the system will not stabilize and subsequently will not converge to its global optimality. If too many evaluations, it is time-consuming, and the system will usually converge too slowly, as the number of iterations to achieve stability might be exponential to the problem size.

Therefore, there is a fine balance between the number of evaluations and solution quality. We can either do many evaluations at a few temperature levels or do few evaluations at many temperature levels. There are two major ways to set the number of iterations: fixed or varied. The first uses a fixed number of iterations at each temperature, while the second intends to increase the number of iterations at lower temperatures so that the local minima can be fully explored.

3.3 SA ALGORITHM

The simulated annealing algorithm can be summarized as the pseudo code shown in Fig. 3.1.

In order to find a suitable starting temperature T_0, we can use any information about the objective function. If we know the maximum change $\max(\Delta f)$ of the objective function, we can use this to estimate an initial temperature T_0 for a given probability p_0. That is

$$T_0 \approx -\frac{\max(\Delta f)}{\ln p_0}.$$

If we do not know the possible maximum change of the objective function, we can use a heuristic approach. We can start evaluations from a very

Simulated Annealing Algorithm

Objective function $f(\boldsymbol{x})$, $\boldsymbol{x} = (x_1, ..., x_p)^T$
Initialize initial temperature T_0 *and initial guess* $\boldsymbol{x}^{(0)}$
Set final temperature T_f *and max number of iterations* N
Define cooling schedule $T \mapsto \alpha T$, $(0 < \alpha < 1)$
while ($T > T_f$ *and* $n < N$)
  Move randomly to new locations: $\boldsymbol{x}_{n+1} = \boldsymbol{x}_n + \epsilon$ *(random walk)*
  Calculate $\Delta f = f_{n+1}(\boldsymbol{x}_{n+1}) - f_n(\boldsymbol{x}_n)$
  Accept the new solution if better
  if *not improved*
    Generate a random number r
    Accept if $p = \exp[-\Delta f/T] > r$
  end if
  Update the best $\boldsymbol{x}_*$ *and* f_*
  $n = n + 1$
end while

Figure 3.1: Simulated annealing algorithm.

high temperature (so that almost all changes are accepted) and reduce the temperature quickly until about 50% or 60% of the worse moves are accepted, and then use this temperature as the new initial temperature T_0 for proper and relatively slow cooling.

For the final temperature, it should be zero in theory so that no worse move can be accepted. However, if $T_f \to 0$, more unnecessary evaluations are needed. In practice, we simply choose a very small value, say, $T_f = 10^{-10} \sim 10^{-5}$, depending on the required quality of the solutions and time constraints.

3.4 UNCONSTRAINED OPTIMIZATION

Based on the guidelines of choosing the important parameters such as the cooling rate, initial and final temperatures, and the balanced number of iterations, we can implement the simulated annealing using both Matlab and Octave.

For Rosenbrock's banana function

$$f(x, y) = (1 - x)^2 + 100(y - x^2)^2,$$

we know that its global minimum $f_* = 0$ occurs at $(1, 1)$ (see Fig. 3.2). This is a standard test function and quite tough for most algorithms. However, by modifying the program given later in the next chapter, we can find this

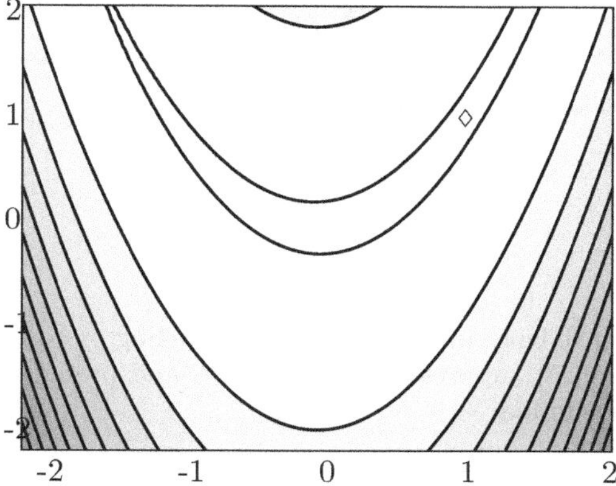

Figure 3.2: Rosenbrock's function with the global minimum $f_* = 0$ at $(1, 1)$.

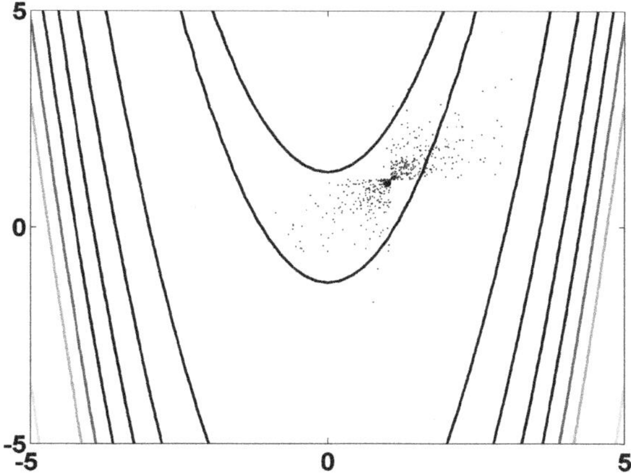

Figure 3.3: 500 evaluations during the annealing iterations. The final global best is marked with •.

global minimum easily and the last 500 evaluations during annealing are shown in Fig. 3.3.

This banana function is still relatively simple as it has a curved narrow valley. We should validate SA against a wide range of test functions, especially those that are strongly multimodal and highly nonlinear. It is straightforward to extend the above program to deal with highly nonlinear multimodal functions.

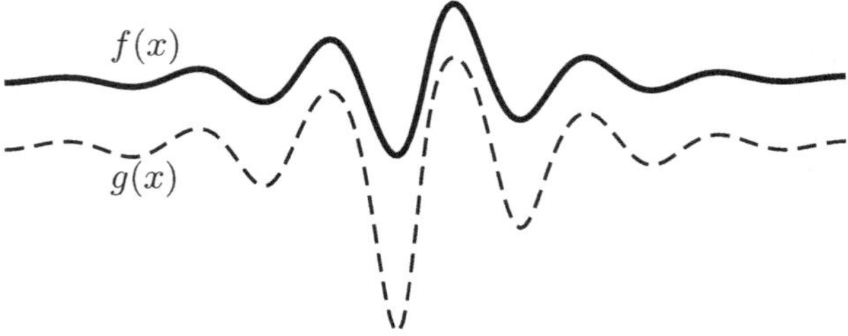

Figure 3.4: The basic idea of stochastic tunneling by transforming $f(x)$ to $g(x)$, suppressing some modes and preserving the locations of minima.

3.5 STOCHASTIC TUNNELING

To ensure the global convergence of simulated annealing, a proper cooling schedule must be used. In the case when the functional landscape is complex, simulated annealing may become increasingly difficult to escape the local minima if the temperature is too low. Raising the temperature, as that in the so-called simulated tempering, may solve the problem, but the convergence is typically slow, and the computing time also increases.

Stochastic tunneling uses the tunneling idea to transform the objective function landscape into a different but more convenient one (e.g., Wenzel and Hamacher, 1999). The essence is to construct a nonlinear transformation so that some modes of $f(x)$ are suppressed and other modes are amplified, while preserving the loci of minima of $f(x)$.

The standard form of such a tunneling transformation is

$$g(x) = 1 - \exp[-\gamma(f(x) - f_0)], \tag{3.7}$$

where f_0 is the current lowest value of $f(x)$ found so far. $\gamma > 0$ is a scaling parameter, and g is the transformed new landscape. From this simple transformation, we can see that $g \to 0$ when $f - f_0 \to 0$, that is when f_0 is approaching the true global minimum. On the other hand, if $f \gg f_0$, then $g \to 1$, which means that all the modes well above the current minimum f_0 are suppressed. For a simple one-dimensional function, it is easy to see that such properties indeed preserve the loci of the function (see Fig. 3.4).

As the loci of the minima are preserved, then all the modes that above the current lowest value f_0 are suppressed to some degree, while the modes below f_0 are expanded or amplified, which makes it easy for the system to escape local modes. Simulations and studies suggest that it can significantly improve the convergence for functions with complex landscape and modes.

Up to now we have not actually provided a detailed program to show how the SA algorithm can be implemented in practice. However, before we can actually do this, we need to find a practical way to deal with constraints, as most real-world optimization problems are constrained. In the next chapter, we will discuss in detail the ways of incorporating nonlinear constraints.

REFERENCES

1. Cerny V., A thermodynamical approach to the travelling salesman problem: an efficient simulation algorithm, *Journal of Optimization Theory and Applications*, **45**, 41-51 (1985).
2. Hamacher K. and Wenzel W., The scaling behaviour of stochastic minimization algorithms in a perfect funnel landscape, *Phys. rev. E.*, **59**, 938-941 (1999).
3. Kirkpatrick S., Gelatt C. D., and Vecchi M. P., Optimization by simulated annealing, *Science*, **220**, No. 4598, 671-680 (1983).
4. Metropolis N., Rosenbluth A. W., Rosenbluth M. N., Teller A. H., and Teller E., Equations of state calculations by fast computing machines, *Journal of Chemical Physics*, **21**, 1087-1092 (1953).
5. Wenzel W. and Hamacher K., A stochastic tunneling approach for global optimization, *Phys. Rev. Lett.*, **82**, 3003-3007 (1999).
6. Yang X. S., Biology-derived algorithms in engineering optimization (Chapter 32), in *Handbook of Bioinspired Algorithms*, edited by Olariu S. and Zomaya A., Chapman & Hall / CRC, (2005).

Chapter 4

HOW TO DEAL WITH CONSTRAINTS

The optimization we have discussed so far is unconstrained, as we have not considered any constraints. A natural and important question is how to incorporate the constraints (both inequality and equality constraints). There are mainly three ways to deal with constraints: direct approach, Lagrange multipliers, and penalty method.

Direct approach intends to find the feasible regions enclosed by the constraints. This is often difficult, except for a few special cases. Numerically, we can generate a potential solution, and check if all the constraints are satisfied. If all the constraints are met, then it is a feasible solution, and the evaluation of the objective function can be carried out. If one or more constraints are not satisfied, this potential solution is discarded, and a new solution should be generated. We then proceed in a similar manner. As we can expect, this process is slow and inefficient. A better approach is to incorporate the constraints so as to formulate the problem as an unconstrained one. The method of Lagrange multiplier has rigorous mathematical basis, while the penalty method is simple to implement in practice.

4.1 METHOD OF LAGRANGE MULTIPLIERS

The method of Lagrange multipliers converts a constrained problem to an unconstrained one. For example, if we want to minimize a function

$$\underset{\boldsymbol{x}\in\Re^n}{\text{minimize}}\, f(\boldsymbol{x}), \qquad \boldsymbol{x}=(x_1,...,x_n)^T\in\Re^n, \tag{4.1}$$

subject to multiple nonlinear equality constraints

$$g_j(\boldsymbol{x})=0, \qquad j=1,2,...,M. \tag{4.2}$$

We can use M Lagrange multipliers $\lambda_j (j=1,...,M)$ to reformulate the above problem as the minimization of the following function

$$L(\boldsymbol{x},\lambda_j)=f(\boldsymbol{x})+\sum_{j=1}^{M}\lambda_j g_j(\boldsymbol{x}). \tag{4.3}$$

The optimality requires that the following stationary conditions hold

$$\frac{\partial L}{\partial x_i} = \frac{\partial f}{\partial x_i} + \sum_{j=1}^{M} \lambda_j \frac{\partial g_j}{\partial x_i}, \quad (i = 1, ..., n), \tag{4.4}$$

and

$$\frac{\partial L}{\partial \lambda_j} = g_j = 0, \quad (j = 1, ..., M). \tag{4.5}$$

These $M + n$ equations will determine the n components of $\boldsymbol{x}$ and M Lagrange multipliers. As $\frac{\partial L}{\partial g_j} = \lambda_j$, we can consider λ_j as the rate of the change of the quantity $L(\boldsymbol{x}, \lambda_j)$ as a functional of g_j.

Now let us look at a simple example

$$\underset{u,v}{\text{maximize}} \; f = u^{2/3} v^{1/3},$$

subject to

$$3u + v = 9.$$

First, we write it as an unconstrained problem using a Lagrange multiplier λ, and we have

$$L = u^{2/3} v^{1/3} + \lambda(3u + v - 9).$$

The conditions to reach optimality are

$$\frac{\partial L}{\partial u} = \frac{2}{3} u^{-1/3} v^{1/3} + 3\lambda = 0, \qquad \frac{\partial L}{\partial v} = \frac{1}{3} u^{2/3} v^{-2/3} + \lambda = 0,$$

and

$$\frac{\partial L}{\partial \lambda} = 3u + v - 9 = 0.$$

The first two conditions give $2v = 3u$, whose combination with the third condition leads to

$$u = 2, \qquad v = 3.$$

Thus, the maximum of f_* is $\sqrt[3]{12}$.

Here we only discussed the equality constraints. For inequality constraints, things become more complicated. We need the so-called Karush-Kuhn-Tucker conditions.

Let us consider the following, generic, nonlinear optimization problem

$$\underset{\boldsymbol{x} \in \Re^n}{\text{minimize}} \; f(\boldsymbol{x}),$$

$$\text{subject to } \phi_i(\boldsymbol{x}) = 0, \quad (i = 1, ..., M),$$

$$\psi_j(\boldsymbol{x}) \leq 0, \quad (j = 1, ..., N). \tag{4.6}$$

If all the functions are continuously differentiable, at a local minimum $\boldsymbol{x}_*$, there exist constants $\lambda_1, ..., \lambda_M$ and $\mu_0, \mu_1, ..., \mu_N$ such that the following KKT optimality conditions hold

$$\mu_0 \nabla f(\boldsymbol{x}_*) + \sum_{i=1}^{M} \lambda_i \nabla \phi_i(\boldsymbol{x}_*) + \sum_{j=1}^{N} \mu_j \nabla \psi_j(\boldsymbol{x}_*) = 0, \tag{4.7}$$

and

$$\psi_j(\boldsymbol{x}_*) \leq 0, \qquad \mu_j \psi_j(\boldsymbol{x}_*) = 0, \ \ (j = 1, 2, ..., N), \tag{4.8}$$

where

$$\mu_j \geq 0, \ \ (j = 0, 1, ..., N). \tag{4.9}$$

The last non-negativity conditions hold for all μ_j, though there is no constraint on the sign of λ_i.

The constants satisfy the following condition

$$\sum_{j=0}^{N} \mu_j + \sum_{i=1}^{M} |\lambda_i| \geq 0. \tag{4.10}$$

This is essentially a generalized method of Lagrange multipliers. However, there is a possibility of degeneracy when $\mu_0 = 0$ under certain conditions. There are two possibilities: 1) there exist vectors $\boldsymbol{\lambda}^* = (\lambda_1^*, ..., \lambda_M^*)^T$ and $\boldsymbol{\mu}^* = (\mu_1^*, .., \mu_N^*)^T$ such that above equations hold, or 2) all the vectors $\nabla \phi_1(\boldsymbol{x}_*), \nabla \phi_2(\boldsymbol{x}_*), ..., \nabla \psi_1(\boldsymbol{x}_*), ..., \nabla \psi_N(\boldsymbol{x}_*)$ are linearly independent, and in this case, the stationary conditions $\frac{\partial L}{\partial x_i}$ do not necessarily hold. As the second case is a special case, we will not discuss this further.

The condition $\mu_j \psi_j(\boldsymbol{x}_*) = 0$ in (4.8) is often called the complementarity condition or complementary slackness condition. It either means $\mu_j = 0$ or $\psi_j(\boldsymbol{x}_*) = 0$. The later case $\psi_j(\boldsymbol{x}_*) = 0$ for any particular j means the inequality becomes tight, and thus becoming an equality. For the former case $\mu_j = 0$, the inequality for a particular j holds and is not tight; however, $\mu_j = 0$ means that this corresponding inequality can be ignored. Therefore, those inequalities that are not tight are ignored, while inequalities which are tight become equalities; consequently, the constrained problem with equality and inequality constraints now essentially becomes a modified constrained problem with selected equality constraints. This is the beauty of the KKT conditions. The main issue remains to identify which inequality becomes tight, and this depends on the individual optimization problem.

The KKT conditions form the basis for mathematical analysis of nonlinear optimization problems, but the numerical implementation of these conditions is not easy, and often inefficient. From the numerical point of view, the penalty method is more straightforward to implement.

4.2 PENALTY METHOD

For a nonlinear optimization problem with equality and inequality constraints, a common method of incorporating constraints is the penalty method. For the optimization problem

$$\underset{\boldsymbol{x}\in\Re^n}{\text{minimize}} f(\boldsymbol{x}), \quad \boldsymbol{x} = (x_1, ..., x_n)^T \in \Re^n,$$

$$\text{subject to } \phi_i(\boldsymbol{x}) = 0, \ (i = 1, ..., M),$$

$$\psi_j(\boldsymbol{x}) \leq 0, \ (j = 1, ..., N), \tag{4.11}$$

the idea is to define a penalty function so that the constrained problem is transformed into an unconstrained problem. Now we define

$$\Pi(\boldsymbol{x}, \mu_i, \nu_j) = f(\boldsymbol{x}) + \sum_{i=1}^{M} \mu_i \phi_i^2(\boldsymbol{x}) + \sum_{j=1}^{N} \nu_j \psi_j^2(\boldsymbol{x}), \tag{4.12}$$

where $\mu_i \gg 1$ and $\nu_j \geq 0$ which should be large enough, depending on the solution quality needed.

As we can see, when an equality constraint it met, its effect or contribution to Π is zero. However, when it is violated, it is penalized heavily as it increases Π significantly. Similarly, it is true when inequality constraints become tight or exact. For the ease of numerical implementation, we should use index functions H to rewrite above penalty function as

$$\Pi = f(\boldsymbol{x}) + \sum_{i=1}^{M} \mu_i H_i[\phi_i(\boldsymbol{x})] \phi_i^2(\boldsymbol{x}) + \sum_{j=1}^{N} \nu_j H_j[\psi_j(\boldsymbol{x})] \psi_j^2(\boldsymbol{x}), \tag{4.13}$$

Here $H_i[\phi_i(\boldsymbol{x})]$ and $H_j[\psi_j(\boldsymbol{x})]$ are index functions.

More specifically, $H_i[\phi_i(\boldsymbol{x})] = 1$ if $\phi_i(\boldsymbol{x}) \neq 0$, and $H_i = 0$ if $\phi_i(\boldsymbol{x}) = 0$. Similarly, $H_j[\psi_j(\boldsymbol{x})] = 0$ if $\psi_j(\boldsymbol{x}) \leq 0$ is true, while $H_j = 1$ if $\psi_j(\boldsymbol{x}) > 0$. In principle, the numerical accuracy depends on the values of μ_i and ν_j which should be reasonably large. But how large is large enough? As most computers have a machine precision of $\epsilon = 2^{-52} \approx 2.2 \times 10^{-16}$, μ_i and ν_j should be close to the order of 10^{15}. Obviously, it could cause numerical problems if they are too large.

In addition, for simplicity of implementation, we can use $\mu = \mu_i$ for all i and $\nu = \nu_j$ for all j. That is, we can use a simplified

$$\Pi(\boldsymbol{x}, \mu, \nu) = f(\boldsymbol{x}) + \mu \sum_{i=1}^{M} H_i[\phi_i(\boldsymbol{x})] \phi_i^2(\boldsymbol{x}) + \nu \sum_{j=1}^{N} H_j[\psi_j(\boldsymbol{x})] \psi_j^2(\boldsymbol{x}).$$

In general, for most applications, μ and ν can be taken as 10^{10} to 10^{15}. We will use these values in our implementation.

Sometimes, it might be easier to change an equality constraint to two inequality constraints, so that we only have to deal with inequalities in the implementation. This is because $g(\boldsymbol{x}) = 0$ is always equivalent to $g(\boldsymbol{x}) \leq 0$ and $g(\boldsymbol{x}) \geq 0$ (or $-g(\boldsymbol{x}) \leq 0$).

4.3 STEP SIZE IN RANDOM WALKS

As random walks are widely used for randomization and local search, a proper step size is very important. In the generic equation

$$\boldsymbol{x}^{t+1} = \boldsymbol{x}^t + s\,\epsilon_t, \tag{4.14}$$

ϵ_t is drawn from a standard normal distribution with zero mean and unity standard deviation. Here the step size s determines how far a random walker (e.g., an agent or particle in metaheuristics) can go for a fixed number of iterations.

If s is too large, then the new solution $\boldsymbol{x}^{t+1}$ generated will be too far away from the old solution (or more often the current best). Then, such a move is unlikely to be accepted. If s is too small, the change is too small to be significant, and consequently such search is not efficient. So a proper step size is important to maintain the search as efficient as possible.

From the theory of simple isotropic random walks, we know that the average distance r traveled in the d-dimension space is

$$r^2 = 2dDt, \tag{4.15}$$

where $D = s^2/2\tau$ is the effective diffusion coefficient. Here s is the step size or distance traveled at each jump, and τ is the time taken for each jump. The above equation implies that

$$s^2 = \frac{\tau\, r^2}{t\, d}. \tag{4.16}$$

For a typical length scale L of a dimension of interest, the local search is typically limited in a region of $L/10$. That is, $r = L/10$. As the iterations are discrete, we can take $\tau = 1$. Typically in metaheuristics, we can expect that the number of generations is usually $t = 100$ to 1000, which means that

$$s \approx \frac{r}{\sqrt{td}} = \frac{L/10}{\sqrt{t\, d}}. \tag{4.17}$$

For $d = 1$ and $t = 100$, we have $s = 0.01L$, while $s = 0.001L$ for $d = 10$ and $t = 1000$. As step sizes could differ from variable to variable, a step size ratio s/L is more generic. Therefore, we can use $s/L = 0.001$ to 0.01 for most problems. We will use this step size factor in our implementation, to be discussed later in the last section of this chapter.

In order to demonstrate the way we incorporate the constraints and the way to do the random walk, it is easy to illustrate using a real-world design example in engineering applications. Now let us look at the well-known welded beam design.

4.4 WELDED BEAM DESIGN

The welded beam design problem is a standard test problem for constrained design optimization, which was described in detail in the literature (Ragsdell and Phillips 1976, Cagnina et al 2008). The problem has four design variables: the width w and length L of the welded area, the depth d and thickness h of the beam. The objective is to minimize the overall fabrication cost, under the appropriate constraints of shear stress τ, bending stress σ, buckling load P and end deflection δ.

The problem can be written as

$$\text{minimize} \quad f(\boldsymbol{x}) = 1.10471 w^2 L + 0.04811 dh(14.0 + L), \tag{4.18}$$

subject to

$$\begin{aligned} g_1(\boldsymbol{x}) &= \tau(\boldsymbol{x}) - 13,600 \le 0 \\ g_2(\boldsymbol{x}) &= \sigma(\boldsymbol{x}) - 30,000 \le 0 \\ g_3(\boldsymbol{x}) &= w - h \le 0 \\ g_4(\boldsymbol{x}) &= 0.10471 w^2 + 0.04811 hd(14 + L) - 5.0 \le 0 \\ g_5(\boldsymbol{x}) &= 0.125 - w \le 0 \\ g_6(\boldsymbol{x}) &= \delta(\boldsymbol{x}) - 0.25 \le 0 \\ g_7(\boldsymbol{x}) &= 6000 - P(\boldsymbol{x}) \le 0, \end{aligned} \tag{4.19}$$

where

$$\sigma(\boldsymbol{x}) = \frac{504,000}{hd^2}, \quad \delta = \frac{65,856}{30,000 hd^3}, \quad Q = 6000(14 + \frac{L}{2}),$$

$$D = \frac{1}{2}\sqrt{L^2 + (w+d)^2}, \quad J = \sqrt{2}\, wL[\frac{L^2}{6} + \frac{(w+d)^2}{2}], \quad \beta = \frac{QD}{J},$$

$$\alpha = \frac{6000}{\sqrt{2} wL}, \quad \tau(\boldsymbol{x}) = \sqrt{\alpha^2 + \frac{\alpha\beta L}{D} + \beta^2},$$

$$P = 0.61423 \times 10^6 \, \frac{dh^3}{6}(1 - \frac{d\sqrt{30/48}}{28}). \tag{4.20}$$

The simple limits or bounds are $0.1 \le L, d \le 10$ and $0.1 \le w, h \le 2.0$.

If we use the simulated annealing algorithm to solve this problem (see next section), we can get the optimal solution which is about the same solution obtained by Cagnina et al (2008)

$$f_* = 1.724852 \text{ at } \quad (0.205730, 3.470489, 9.036624, 0.205729). \tag{4.21}$$

It is worth pointing out that you have to run the programs a few times using values such as $\alpha = 0.95$ (default) and $\alpha = 0.99$ to see how the results vary. In addition, as SA is a stochastic optimization algorithm, we cannot expect the results are the same. In fact, they will be slightly different, every time we run the program. Therefore, we should understand and interpret the results using statistical measures such as mean and standard deviation.

4.5 SA IMPLEMENTATION

We just formulated the welded beam design problem using different notations from some literature. Here we try to illustrate a point.

As the input to a function is a vector (either column vector or less often row vector), we have to write

$$\boldsymbol{x} = \begin{pmatrix} w & L & d & h \end{pmatrix} = [x(1) \;\; x(2) \;\; x(3) \;\; x(4)]. \tag{4.22}$$

With this vector, the objective becomes

$$\text{minimize } f(\boldsymbol{x}) = 1.10471 * x(1)^2 * x(2) + 0.04811 * x(3) * x(4)(14.0 + x(2)),$$

which can easily be converted to a formula in Matlab. Similarly, the third inequality constraint can be rewritten as

$$g_3 = g(3) = x(1) - x(4) \le 0. \tag{4.23}$$

Other constraints can be rewritten in a similar way.

Using the pseudo code for simulated annealing and combining with the penalty method, we can solve the above welded beam design problem using simulated annealing in Matlab as follows:

```
% Simulated Annealing for constrained optimization
% by Xin-She Yang @ Cambridge University @2008
% Usage: sa_mincon(0.99) or sa_mincon;

function [bestsol,fval,N]=sa_mincon(alpha)
% Default cooling factor
if nargin<1,
   alpha=0.95;
end

% Display usage
disp('sa_mincon or [Best,fmin,N]=sa_mincon(0.9)');

% Welded beam design optimization
Lb=[0.1 0.1  0.1 0.1];
```

```
Ub=[2.0 10.0 10.0 2.0];
u0=(Lb+Ub)/2;

if length(Lb) ~=length(Ub),
    disp('Simple bounds/limits are improper!');
    return
end

%% Start of the main program ------------------------
d=length(Lb);       % Dimension of the problem

% Initializing parameters and settings
T_init = 1.0;       % Initial temperature
T_min =  1e-10;     % Finial stopping temperature
F_min = -1e+100;    % Min value of the function
max_rej=500;        % Maximum number of rejections
max_run=150;        % Maximum number of runs
max_accept = 50;    % Maximum number of accept
initial_search=500; % Initial search period
k = 1;              % Boltzmann constant
Enorm=1e-5;         % Energy norm (eg, Enorm=1e-8)

% Initializing the counters i,j etc
i= 0; j = 0; accept = 0; totaleval = 0;
% Initializing various values
T = T_init;
E_init = Fun(u0);
E_old = E_init; E_new=E_old;
best=u0;  % initially guessed values
% Starting the simulated annealing
while ((T > T_min) & (j <= max_rej) & E_new>F_min)
    i = i+1;
    % Check if max numbers of run/accept are met
    if (i >= max_run) | (accept >= max_accept)
        % reset the counters
        i = 1;  accept = 1;
      % Cooling according to a cooling schedule
        T = cooling(alpha,T);
    end

    % Function evaluations at new locations
    if totaleval<initial_search,
        init_flag=1;
        ns=newsolution(u0,Lb,Ub,init_flag);
```

```
    else
        init_flag=0;
        ns=newsolution(best,Lb,Ub,init_flag);
    end
      totaleval=totaleval+1;
      E_new = Fun(ns);
    % Decide to accept the new solution
    DeltaE=E_new-E_old;
    % Accept if improved
    if (DeltaE <0)
        best = ns; E_old = E_new;
        accept=accept+1;   j = 0;
    end
    % Accept with a probability if not improved
    if (DeltaE>=0 & exp(-DeltaE/(k*T))>rand );
        best = ns; E_old = E_new;
        accept=accept+1;
    else
        j=j+1;
    end
    % Update the estimated optimal solution
    f_opt=E_old;
end

bestsol=best;
fval=f_opt;
N=totaleval;

%% New solutions
function s=newsolution(u0,Lb,Ub,init_flag)
  % Either search around
if length(Lb)>0 & init_flag==1,
  s=Lb+(Ub-Lb).*rand(size(u0));
else
  % Or local search by random walk
  stepsize=0.01;
  s=u0+stepsize*(Ub-Lb).*randn(size(u0));
end

s=bounds(s,Lb,Ub);

%% Cooling
function T=cooling(alpha,T)
T=alpha*T;
```

```
function ns=bounds(ns,Lb,Ub)
if length(Lb)>0,
% Apply the lower bound
  ns_tmp=ns;
  I=ns_tmp<Lb;
  ns_tmp(I)=Lb(I);
  % Apply the upper bounds
  J=ns_tmp>Ub;
  ns_tmp(J)=Ub(J);
% Update this new move
  ns=ns_tmp;
else
  ns=ns;
end

% d-dimensional objective function
function z=Fun(u)

% Objective
z=fobj(u);

% Apply nonlinear constraints by penalty method
% Z=f+sum_k=1^N lam_k g_k^2 *H(g_k)
z=z+getnonlinear(u);

function Z=getnonlinear(u)
Z=0;
% Penalty constant
lam=10^15; lameq=10^15;
[g,geq]=constraints(u);

% Inequality constraints
for k=1:length(g),
    Z=Z+ lam*g(k)^2*getH(g(k));
end

% Equality constraints (when geq=[], length->0)
for k=1:length(geq),
   Z=Z+lameq*geq(k)^2*geteqH(geq(k));
end

% Test if inequalities hold
function H=getH(g)
```

```
if g<=0,
    H=0;
else
    H=1;
end

% Test if equalities hold
function H=geteqH(g)
if g==0,
    H=0;
else
    H=1;
end

% Objective functions
function z=fobj(u)
% Welded beam design optimization
z=1.10471*u(1)^2*u(2)+0.04811*u(3)*u(4)*(14.0+u(2));

% All constraints
function [g,geq]=constraints(x)
% Inequality constraints
Q=6000*(14+x(2)/2);
D=sqrt(x(2)^2/4+(x(1)+x(3))^2/4);
J=2*(x(1)*x(2)*sqrt(2)*(x(2)^2/12+(x(1)+x(3))^2/4));
alpha=6000/(sqrt(2)*x(1)*x(2));
beta=Q*D/J;
tau=sqrt(alpha^2+2*alpha*beta*x(2)/(2*D)+beta^2);
sigma=504000/(x(4)*x(3)^2);
delta=65856000/(30*10^6*x(4)*x(3)^3);
tmpf=4.013*(30*10^6)/196;
P=tmpf*sqrt(x(3)^2*x(4)^6/36)*(1-x(3)*sqrt(30/48)/28);

g(1)=tau-13600;
g(2)=sigma-30000;
g(3)=x(1)-x(4);
g(4)=0.10471*x(1)^2+0.04811*x(3)*x(4)*(14+x(2))-5.0;
g(5)=0.125-x(1);
g(6)=delta-0.25;
g(7)=6000-P;

% Equality constraints
geq=[];
%% End of the program --------------------------------
```

How to Get the Files

To get the files of all the Matlab programs provided in this book, readers can send an email (with the subject 'Nature-Inspired Algorithms: Files') to `Metaheuristic.Algorithms@gmail.com` – A zip file will be provided (via email) by the author.

REFERENCES

1. Cagnina L. C., Esquivel S. C., and Coello C. A., Solving engineering optimization problems with the simple constrained particle swarm optimizer, *Informatica*, **32**, 319-326 (2008)
2. Cerny V., A thermodynamical approach to the travelling salesman problem: an efficient simulation algorithm, *Journal of Optimization Theory and Applications*, **45**, 41-51 (1985).
3. Deb K., *Optimisation for Engineering Design*: Algorithms and Examples, Prentice-Hall, New Delhi, (1995).
4. Gill P. E., Murray W., and Wright M. H., *Practical optimization*, Academic Press Inc, (1981).
5. Hamacher K., Wenzel W., The scaling behaviour of stochastic minimization algorithms in a perfect funnel landscape, *Phys. Rev. E.*, **59**, 938-941(1999).
6. Kirkpatrick S., Gelatt C. D., and Vecchi M. P., Optimization by simulated annealing, *Science*, **220**, No. 4598, 671-680 (1983).
7. Metropolis N., Rosenbluth A. W., Rosenbluth M. N., Teller A. H., and Teller E., Equations of state calculations by fast computing machines, *Journal of Chemical Physics*, **21**, 1087-1092 (1953).
8. Ragsdell K. and Phillips D., Optimal design of a class of welded structures using geometric programming, *J. Eng. Ind.*, **98**, 1021-1025 (1976).
9. Wenzel W. and Hamacher K., A stochastic tunneling approach for global optimization, *Phys. Rev. Lett.*, **82**, 3003-3007 (1999).
10. Yang X. S., Biology-derived algorithms in engineering optimization (Chapter 32), in *Handbook of Bioinspired Algorithms*, edited by Olariu S. and Zomaya A., Chapman & Hall / CRC, (2005).
11. E. G. Talbi, *Metaheuristics: From Design to Implementation*, Wiley, (2009).

Chapter 5

GENETIC ALGORITHMS

Genetic algorithms are probably the most popular evolutionary algorithms in terms of the diversity of their applications. A vast majority of well-known optimization problems have been tried by genetic algorithms. In addition, genetic algorithms are population-based, and many modern evolutionary algorithms are directly based on genetic algorithms, or have some strong similarities.

5.1 INTRODUCTION

The genetic algorithm (GA), developed by John Holland and his collaborators in the 1960s and 1970s, is a model or abstraction of biological evolution based on Charles Darwin's theory of natural selection. Holland was the first to use the crossover and recombination, mutation, and selection in the study of adaptive and artificial systems. These genetic operators form the essential part of the genetic algorithm as a problem-solving strategy. Since then, many variants of genetic algorithms have been developed and applied to a wide range of optimization problems, from graph colouring to pattern recognition, from discrete systems (such as the travelling salesman problem) to continuous systems (e.g., the efficient design of airfoil in aerospace engineering), and from financial market to multiobjective engineering optimization.

There are many advantages of genetic algorithms over traditional optimization algorithms, and two most noticeable advantages are: the ability of dealing with complex problems and parallelism. Genetic algorithms can deal with various types of optimization whether the objective (fitness) function is stationary or non-stationary (change with time), linear or nonlinear, continuous or discontinuous, or with random noise. As multiple offsprings in a population act like independent agents, the population (or any subgroup) can explore the search space in many directions simultaneously. This feature makes it ideal to parallelize the algorithms for implementation. Different parameters and even different groups of encoded strings can be manipulated at the same time.

Genetic Algorithm

Objective function $f(\boldsymbol{x})$, $\boldsymbol{x} = (x_1, ..., x_n)^T$
Encode the solution into chromosomes (binary strings)
Define fitness F *(eg,* $F \propto f(\boldsymbol{x})$ *for maximization)*
Generate the initial population
Initial probabilities of crossover (p_c) *and mutation* (p_m)
  while (t <*Max number of generations*)
    Generate new solution by crossover and mutation
    if p_c >*rand*, *Crossover;* **end if**
    if p_m >*rand*, *Mutate;* **end if**
    Accept the new solution if its fitness increases
    Select the current best for the next generation (elitism)
  end while
Decode the results and visualization

Figure 5.1: Pseudo code of genetic algorithms.

However, genetic algorithms also have some minor disadvantages. The formulation of fitness function, the usage of population size, the choice of the important parameters such as the rate of mutation and crossover, and the selection criteria of new population should carefully be carried out. Any inappropriate choice will make it difficult for the algorithm to converge, or it simply produces meaningless results. Despite these, genetic algorithms remain one of the most widely used optimization algorithms in modern nonlinear optimization.

5.2 GENETIC ALGORITHMS

The essence of genetic algorithms involves the encoding of an optimization function as arrays of bits or character strings to represent the chromosomes, the manipulation operations of strings by genetic operators, and the selection according to their fitness with the aim to find a solution to the problem concerned.

This is often done by the following procedure: 1) encoding of the objectives or optimization functions; 2) defining a fitness function or selection criterion; 3) creating a population of individuals; 4) evolution cycle or iterations by evaluating the fitness of all the individuals in the population, creating a new population by performing crossover, and mutation, fitness-proportionate reproduction etc, and replacing the old population and iterating again using the new population; 5) decoding the results to obtain the solution to the problem. These steps can schematically be represented as the pseudo code of genetic algorithms shown in Fig. 5.1.

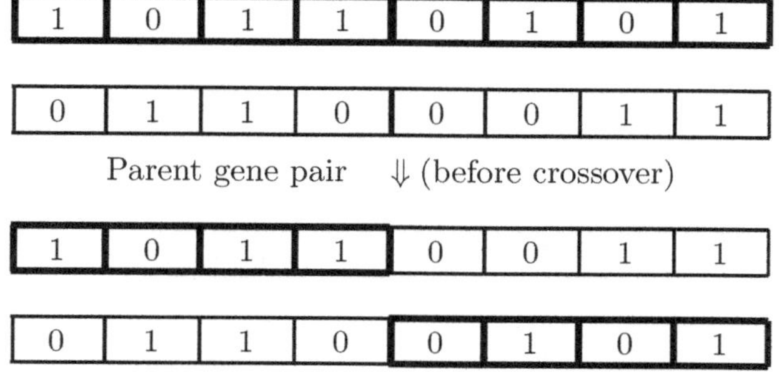

Figure 5.2: Diagram of crossover at a random crossover point (location) in genetic algorithms.

One iteration of creating a new population is called a generation. The fixed-length character strings are used in most of genetic algorithms during each generation although there is substantial research on the variable-length strings and coding structures. The coding of the objective function is usually in the form of binary arrays or real-valued arrays in the adaptive genetic algorithms. For simplicity, we use binary strings for encoding and decoding in our discussion. The genetic operators include crossover, mutation, and selection from the population.

The crossover of two parent strings is the main operator with a higher probability p_c and is carried out by swapping one segment of one chromosome with the corresponding segment on another chromosome at a random position (see Fig. 5.2). The crossover carried out in this way is a single-point crossover. Crossover at multiple points is more often used in genetic algorithms to increase the evolutionary efficiency of the algorithms.

The mutation operation is achieved by flopping the randomly selected bits (see Fig. 5.3), and the mutation probability p_m is usually small.

The selection of an individual in a population is carried out by the evaluation of its fitness, and it can remain in the new generation if a certain threshold of the fitness is reached, we can also use that the reproduction of a population is fitness-proportionate. That is to say, the individuals with higher fitness are more likely to reproduce.

5.3 CHOICE OF PARAMETERS

An important issue is the formulation or choice of an appropriate fitness function that determines the selection criterion in a particular problem. For the minimization of a function using genetic algorithms, one simple way of constructing a fitness function is to use the simplest form $F = A - y$ with A being a large constant (though $A = 0$ will do if the fitness needs

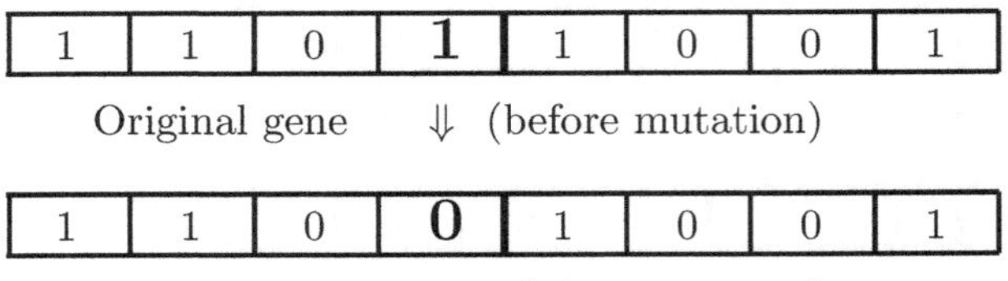

Figure 5.3: Schematic representation of mutation at a single site by flipping a randomly selected bit ($1 \to 0$).

not to be non negative) and $y = f(\mathbf{x})$, thus the objective is to maximize the fitness function and subsequently minimize the objective function $f(\mathbf{x})$. However, there are many different ways of defining a fitness function. For example, we can use the individual fitness assignment relative to the whole population

$$F(x_i) = \frac{f(\xi_i))}{\sum_{i=1}^{N} f(\xi_i)}, \tag{5.1}$$

where ξ_i is the phenotypic value of individual i, and N is the population size. The appropriate form of the fitness function will make sure that the solutions with higher fitness should be selected efficiently. Poor fitness function may result in incorrect or meaningless solutions.

Another important issue is the choice of various parameters. The crossover probability p_c is usually very high, typically in the range of $0.7 \sim 1.0$. On the other hand, the mutation probability p_m is usually small (usually $0.001 \sim 0.05$). If p_c is too small, then the crossover occurs sparsely, which is not efficient for evolution. If the mutation probability is too high, the solutions could still 'jump around' even if the optimal solution is approaching.

A proper criterion for selecting the best solutions is also important. How to select the current population so that the best individuals with higher fitness should be preserved and passed onto the next generation. That is often carried out in association with certain elitism. The basic elitism is to select the most fit individual (in each generation) which will be carried over to the new generation without being modified by genetic operators. This ensures that the best solution is achieved more quickly.

Other issues include the multiple sites for mutation and the use of various population sizes. The mutation at a single site is not very efficient, mutation at multiple sites will increase the evolution efficiency. However, too many mutants will make it difficult for the system to converge or even makes the system go astray to the wrong solutions. In real ecological systems, if the mutation rate is too high under high selection pressure, then the whole population might go extinct.

Figure 5.4: Encode all design variables into a single long string.

In addition, the choice of the right population size is also very important. If the population size is too small, there is not enough evolution going on, and there is a risk for the whole population to go extinct. In the real world, a species with a small population, ecological theory suggests that there is a real danger of extinction for such species. Even the system carries on, there is still a danger of premature convergence. In a small population, if a significantly more fit individual appears too early, it may reproduces enough offsprings so that they overwhelm the whole (small) population. This will eventually drive the system to a local optimum (not the global optimum). On the other hand, if the population is too large, more evaluations of the objective function are needed, which will require extensive computing time.

Furthermore, more complex and adaptive genetic algorithms are under active research and the literature is vast about these topics.

Using the basic procedure described in the above section, we can implement the genetic algorithms in any programming language. From the implementation point of view, there are two main ways to use encoded strings. We can either use a string (or multiple strings) for each individual in a population, or use a long string for all the individuals (or all design variables) shown in Fig. 5.4.

For the well-known Easom function

$$f(x) = -\cos(x)e^{-(x-\pi)^2}, \qquad x \in [-10, 10], \qquad (5.2)$$

it has the global maximum $f_{\max} = 1$ at $x_* = \pi$. The outputs from a typical run are shown in Fig. 5.5 where the top figure shows the variations of the best estimates as they approach to $x_* \to \pi$, while the lower figure shows the variations of the fitness function.

Genetic algorithms are widely used and there are many software packages in almost all programming languages. Efficient implementation can be found in both commercial and free codes. For this reason, we will not provide any implementation here.

REFERENCES

1. De Jong K., *Analysis of the Behaviour of a Class of Genetic Adaptive Systems*, PhD thesis, University of Michigan, Ann Arbor, (1975).

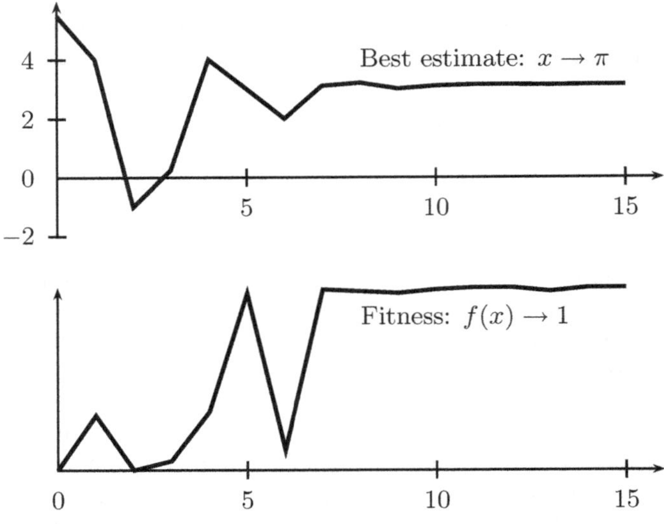

Figure 5.5: Typical outputs from a typical run. The best estimate will approach π while the fitness will approach $f_{\max} = 1$.

2. Fogel D. B., *Evolutionary Computation: Toward a New Philosophy of Machine Intelligence*, IEEE Press, Piscataway, NJ., 3rd Edition, (2006).

3. Goldberg D. E., *Genetic Algorithms in Search, Optimisation and Machine Learning*, Reading, Mass.: Addison Wesley (1989).

4. Holland J., *Adaptation in Natural and Artificial Systems*, University of Michigan Press, Ann Anbor, (1975).

5. Koza J., *Genetic Programming: On the Programming of Computers by Means of Natural Selection*, MIT Press, (1992).

6. Michalewicz Z., *Genetic Algorithms + Data Structures = Evolution Programs*, Springer-Verlag, (1999).

Chapter 6

DIFFERENTIAL EVOLUTION

6.1 INTRODUCTION

Differential evolution (DE) was developed by R. Storn and K. Price by their nominal papers in 1996 and 1997. It is a vector-based evolutionary algorithm, and can be considered as a further development to genetic algorithms. It is a stochastic search algorithm with self-organizing tendency and does not use the information of derivatives. Thus, it is a population-based, derivative-free method. Another advantage of differential evolution over genetic algorithms is that DE treats solutions as real-number strings, thus no encoding and decoding is needed.

As in genetic algorithms, design parameters in a d-dimensional search space are represented as vectors, and various genetic operators are operated over their bits of strings. However, unlikely genetic algorithms, differential evolution carries out operations over each component (or each dimension of the solution). Almost everything is done in terms of vectors. For example, in genetic algorithms, mutation is carried out at one site or multiple sites of a chromosome, while in differential evolution, a difference vector of two randomly-chosen population vectors is used to perturb an existing vector. Such vectorized mutation can be viewed as a self-organizing search, directed towards an optimality. This kind of perturbation is carried out over each population vector, and thus can be expected to be more efficient. Similarly, crossover is also a vector-based component-wise exchange of chromosomes or vector segments.

6.2 DIFFERENTIAL EVOLUTION

For a d-dimensional optimization problem with d parameters, a population of n solution vectors are initially generated, we have $\boldsymbol{x}_i$ where $i = 1, 2, ..., n$. For each solution $\boldsymbol{x}_i$ at any generation t, we use the conventional notation as

$$\boldsymbol{x}_i^t = (x_{1,i}^t, x_{2,i}^t, ..., x_{d,i}^t), \tag{6.1}$$

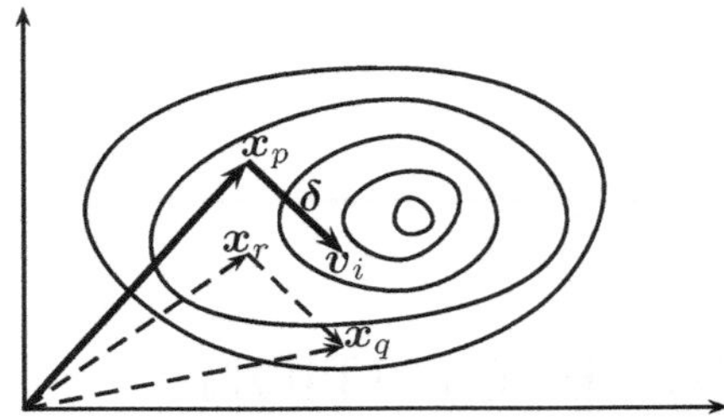

Figure 6.1: Schematic representation of mutation vectors in differential evolution with movement $\boldsymbol{\delta} = F(\boldsymbol{x}_q - \boldsymbol{x}_r)$.

which consists of d-components in the d-dimensional space. This vector can be considered as the chromosomes or genomes.

Differential evolution consists of three main steps: mutation, crossover and selection.

Mutation is carried out by the mutation scheme. For each vector $\boldsymbol{x}_i$ at any time or generation t, we first randomly choose three distinct vectors $\boldsymbol{x}_p$, $\boldsymbol{x}_q$ and $\boldsymbol{x}_r$ at t (see Fig. 6.1), and then generate a so-called donor vector by the mutation scheme

$$\boldsymbol{v}_i^{t+1} = \boldsymbol{x}_p^t + F(\boldsymbol{x}_q^t - \boldsymbol{x}_r^t), \tag{6.2}$$

where $F \in [0, 2]$ is a parameter, often referred to as the differential weight. This requires that the minimum number of population size is $n \geq 4$. In principle, $F \in [0, 2]$, but in practice, a scheme with $F \in [0, 1]$ is more efficient and stable. From Fig. 6.1, we can see that the perturbation $\boldsymbol{\delta} = F(\boldsymbol{x}_q - \boldsymbol{x}_r)$ to the vector $\boldsymbol{x}_p$ is used to generate a donor vector $\boldsymbol{v}_i$, and such perturbation is directed and self-organized.

The crossover is controlled by a crossover probability $C_r \in [0, 1]$ and actual crossover can be carried out in two ways: binomial and exponential. The binomial scheme performs crossover on each of the d components or variables/parameters. By generating a uniformly distributed random number $r_i \in [0, 1]$, the jth component of $\boldsymbol{v}_i$ is manipulated as

$$\boldsymbol{u}_{j,i}^{t+1} = \begin{cases} \boldsymbol{v}_{j,i} & \text{if } r_i \leq C_r \\ \boldsymbol{x}_{j,i}^t & \text{otherwise} \end{cases}, \qquad j = 1, 2, ..., d. \tag{6.3}$$

This way, each component can be decided randomly whether to exchange with donor vector or not.

In the exponential scheme, a segment of the donor vector is selected and this segment starts with a random k with a random length L which can include many components. Mathematically, this is to choose $k \in [0, d-1]$

Differential Evolution

Initialize the population $\boldsymbol{x}$ *with randomly generated solutions*
Set the weight $F \in [0,2]$ *and crossover probability* $C_r \in [0,1]$
while *(stopping criterion)*
  for $i = 1$ *to* n,
  For each $\boldsymbol{x}_i$, *randomly choose 3 distinct vectors* $\boldsymbol{x}_p$, $\boldsymbol{x}_r$ *and* $\boldsymbol{x}_r$
  Generate a new vector $\boldsymbol{v}$ *by DE scheme (6.2)*
  Generate a random index $J_r \in \{1, 2, ..., d\}$ *by permutation*
  Generate a randomly distributed number $r_i \in [0,1]$
  for $j = 1$ *to* d,
  For each parameter $\boldsymbol{v}_{j,i}$ *(jth component of* $\boldsymbol{v}_i$*), update*

$$\boldsymbol{u}_{j,i}^{t+1} = \begin{cases} \boldsymbol{v}_{j,i}^{t+1} & \textit{if } r_i \leq C_r \textit{ or } j = J_r \\ \boldsymbol{x}_{j,i}^{t} & \textit{if } r_i > C_r \textit{ and } j \neq J_r, \end{cases}$$

  end
  Select and update the solution by (6.5)
  end
  Update the counters such as $t = t + 1$
end
Post-process and output the best solution found

Figure 6.2: Pseudo code of differential evolution.

and $L \in [1, d]$ randomly, and we have

$$\boldsymbol{u}_{j,i}^{t+1} = \begin{cases} \boldsymbol{v}_{j,i}^{t} & \text{for } j = k, ..., k - L + 1 \in [1, d] \\ \\ \boldsymbol{x}_{j,i}^{t} & \text{otherwise.} \end{cases} \tag{6.4}$$

As the binomial is simpler to implement, we will use the binomial crossover in our implementation.

Selection is essentially the same as that used in genetic algorithms. It is to select the most fittest, and for minimization problem, the minimum objective value. Therefore, we have

$$\boldsymbol{x}_i^{t+1} = \begin{cases} \boldsymbol{u}_i^{t+1} & \text{if } f(\boldsymbol{u}_i^{t+1}) \leq f(\boldsymbol{x}_i^t), \\ \boldsymbol{x}_i^t & \text{otherwise.} \end{cases} \tag{6.5}$$

All the above three components can be seen in the pseudo code as shown in Fig. 6.2. It is worth pointing out here that the use of J is to ensure that $\boldsymbol{v}_i^{t+1} \neq \boldsymbol{x}_i^t$, which may increase the evolutionary or exploratory efficiency. The overall search efficiency is controlled by two parameters: the differential weight F and the crossover probability C_r.

6.3 VARIANTS

Most studies have focused on the choice of F, C_r and n as well as the modification of (6.2). In fact, when generating mutation vectors, we can use many different ways of formulating (6.2), and this leads to various schemes with the naming convention: DE/x/y/z where x is the mutation scheme (rand or best), y is the number of difference vectors, and z is the crossover scheme (binomial or exponential). The basic DE/Rand/1/Bin scheme is given in (6.2), that is

$$\boldsymbol{v}_i^{t+1} = \boldsymbol{x}_p^t + F(\boldsymbol{x}_q^t - \boldsymbol{x}_r^t). \tag{6.6}$$

If we replace the $\boldsymbol{x}_p^t$ by the current best $\boldsymbol{x}_{\text{best}}$ found so far, we have the so-called DE/Best/1/Bin scheme

$$\boldsymbol{v}_i^{t+1} = \boldsymbol{x}_{\text{best}}^t + F(\boldsymbol{x}_q^t - \boldsymbol{x}_r^t). \tag{6.7}$$

There is no reason that why we should not use more than 3 distinct vectors. For example. if we use 4 different vectors plus the current best, we have the DE/Best/2/Bin scheme

$$\boldsymbol{v}_i^{t+1} = \boldsymbol{x}_{\text{best}}^t + F(\boldsymbol{x}_{k_1}^t + \boldsymbol{x}_{k_2}^t - \boldsymbol{x}_{k_3}^t - \boldsymbol{x}_{k_4}^t). \tag{6.8}$$

Furthermore, if we use five different vectors, we have the DE/Rand/2/Bin scheme

$$\boldsymbol{v}_i^{t+1} = \boldsymbol{x}_{k_1}^t + F_1(\boldsymbol{x}_{k_2}^t - \boldsymbol{x}_{k_3}^t) + F_2(\boldsymbol{x}_{k_4}^t - \boldsymbol{x}_{k_5}^t), \tag{6.9}$$

where F_1 and F_2 are differential weights in $[0, 1]$. Obviously, for simplicity, we can also take $F_1 = F_2 = F$.

Following the similar strategy, we can design various schemes. In fact, 10 different schemes have been formulated, and for details, readers can refer to Price, Storn and Lampinin (2005).

6.4 IMPLEMENTATION

The implementation of differential evolution is relatively straightforward, comparing with genetic algorithms. If a vector/matrix-based software package such as Matlab is used, the implementation becomes even more simple.

A simple version of the DE in Matlab/Octave is given below, which is for unconstrained optimization where Rosenbrock's function with three variables is solved by default. To solve constrained optimization problems, this program can easily be extended in combination with the penalty method discussed earlier when we solved the welded beam design problem.

```
% Differential Evolution for global optimization
% Programmed by Xin-She Yang @Cambridge University 2008
```

```
% The basic version of scheme DE/Rand/1 is implemented
% Usage: de(para) or de;

function [best,fmin,N_iter]=de(para)
% Default parameters
if nargin<1,
   para=[10 0.7 0.9];
end

n=para(1);        % Population >=4, typically 10 to 25
F=para(2);        % DE parameter - scaling (0.5 to 0.9)
Cr=para(3);       % DE parameter - crossover probability

% Iteration parameters
tol=10^(-5);      % Stop tolerance
N_iter=0;         % Total number of function evaluations

% Simple bounds
Lb=[-1 -1 -1];
Ub=[2 2  2];

% Dimension of the search variables
d=length(Lb);

% Initialize the population/solutions
for i=1:n,
  Sol(i,:)=Lb+(Ub-Lb).*rand(size(Lb));
  Fitness(i)=Fun(Sol(i,:));
end

% Find the current best
[fmin,I]=min(Fitness);
best=Sol(I,:)

% Start the iterations by differential evolution
while (fmin>tol)
    % Obtain donor vectors by permutation
    k1=randperm(n);     k2=randperm(n);
    k1sol=Sol(k1,:);    k2sol=Sol(k2,:);
        % Random crossover index/matrix
        K=rand(n,d)<Cr;
        % DE/RAND/1 scheme
        V=Sol+F*(k1sol-k2sol);
```

```
        V=Sol.*(1-K)+V.*K;

        % Evaluate new solutions
        for i=1:n,
           Fnew=Fun(V(i,:));
           % If the solution improves
           if Fnew<=Fitness(i),
                Sol(i,:)=V(i,:);
                Fitness(i)=Fnew;
           end
          % Update the current best
          if Fnew<=fmin,
                best=V(i,:);
                fmin=Fnew;
          end
        end
        N_iter=N_iter+n;
end

% Output/display
disp(['Number of evaluations: ',num2str(N_iter)]);
disp(['Best=',num2str(best),' fmin=',num2str(fmin)]);


% Objective function -- Rosenbrock's 3D function
function z=Fun(u)
z=(1-u(1))^2+100*(u(2)-u(1)^2)^2+100*(u(3)-u(2)^2)^2;
```

REFERENCES

1. Mitchell M., *An Introduction to Genetic Algorithms*, MIT Press, (1996).
2. Storn R., On the usage of differential evolution for function optimization, Biennial Conference of the North American Fuzzy Information Processing Society (NAFIPS), pp. 519-523 (1996).
3. Storn R., web pages on differential evolution with various programming codes, http://www.icsi.berkeley.edu/~storn/code.html
4. Storn R. and Price K., Differential evolution - a simple and efficient heuristic for global optimization over continuous spaces, *Journal of Global Optimization*, **11**, 341-359 (1997).
5. Price K., Storn R. and Lampinen J., *Differential Evolution: A Practical Approach to Global Optimization*, Springer, (2005).

Chapter 7

ANT AND BEE ALGORITHMS

From the discussion of genetic algorithms, we know that we can improve the search efficiency by using randomness which will also increase the diversity of the solutions so as to avoid being trapped in local optima. The selection of the best individuals is also equivalent to use memory. In fact, there are other forms of selection such as using chemical messenger (pheromone) which is commonly used by ants, honey bees, and many other insects. In this chapter, we will discuss the nature-inspired ant algorithms and bee algorithms.

7.1 ANT ALGORITHMS

7.1.1 Behaviour of Ants

Ants are social insects in habit and they live together in organized colonies whose population size can range from about 2 to 25 millions. When foraging, a swarm of ants or mobile agents interact or communicate in their local environment. Each ant can lay scent chemicals or pheromone so as to communicate with others, and each ant is also able to follow the route marked with pheromone laid by other ants. When ants find a food source, they will mark it with pheromone and also mark the trails to and from it. From the initial random foraging route, the pheromone concentration varies and the ants follow the route with higher pheromone concentration, and the pheromone is enhanced by the increasing number of ants. As more and more ants follow the same route, it becomes the favored path. Thus, some favorite routes emerge, often the shortest or more efficient. This is actually a positive feedback mechanism.

Emerging behavior exists in an ant colony and such emergence arises from simple interactions among individual ants. Individual ants act according to simple and local information (such as pheromone concentration) to carry out their activities. Although there is no master ant overseeing the entire colony and broadcasting instructions to the individual ants, organized behavior still emerges automatically. Therefore, such emergent behavior is similar to other self-organized phenomena which occur in many processes in nature such as the pattern formation in animal skins (tiger and zebra skins).

The foraging pattern of some ant species (such as the army ants) can show extraordinary regularity. Army ants search for food along some regular routes with an angle of about 123° apart. We do not know how they manage to follow such regularity, but studies show that they could move in an area and build a bivouac and start foraging. On the first day, they forage in a random direction, say, the north and travel a few hundred meters, then branch to cover a large area. The next day, they will choose a different direction, which is about 123° from the direction on the previous day and cover a large area. On the following day, they again choose a different direction about 123° from the second day's direction. In this way, they cover the whole area over about 2 weeks and they move out to a different location to build a bivouac and forage in the new region.

The interesting thing is that they do not use the angle of $360°/3 = 120°$ (this would mean that on the fourth day, they will search on the empty area already foraged on the first day). The beauty of this 123° angle is that it leaves an angle of about 10° from the direction on the first day. This means they cover the whole circular region in 14 days without repeating (or covering a previously-foraged area). This is an amazing phenomenon.

7.1.2 Ant Colony Optimization

Based on these characteristics of ant behaviour, scientists have developed a number of powerful ant colony algorithms with important progress made in recent years. Marco Dorigo pioneered the research in this area in 1992. Many different variants appear since then.

If we only use some of the nature or the behavior of ants and add some new characteristics, we can devise a class of new algorithms. The basic steps of the ant colony optimization (ACO) can be summarized as the pseudo code shown in Fig. 7.1.

There are two important issues here: the probability of choosing a route, and the evaporation rate of pheromone. There are a few ways of solving these problems, although it is still an area of active research. Here we introduce the current best method.

For a network routing problem, the probability of ants at a particular node i to choose the route from node i to node j is given by

$$p_{ij} = \frac{\phi_{ij}^{\alpha} d_{ij}^{\beta}}{\sum_{i,j=1}^{n} \phi_{ij}^{\alpha} d_{ij}^{\beta}}, \tag{7.1}$$

where $\alpha > 0$ and $\beta > 0$ are the influence parameters, and their typical values are $\alpha \approx \beta \approx 2$. ϕ_{ij} is the pheromone concentration on the route between i and j, and d_{ij} the desirability of the same route. Some *a priori* knowledge about the route such as the distance s_{ij} is often used so that

Ant Colony Optimization

Objective function $f(\boldsymbol{x})$, $\boldsymbol{x} = (x_1, ..., x_n)^T$
[or $f(\boldsymbol{x}_{ij})$ *for a routing problem where* $(i,j) \in \{1, 2, ..., n\}$*]*
Define pheromone evaporation rate γ
while *(criterion)*
for *loop over all n dimensions (or nodes)*
Generate new solutions
Evaluate the new solutions
Mark better locations/routes with pheromone $\delta\phi_{ij}$
Update pheromone: $\phi_{ij} \leftarrow (1-\gamma)\phi_{ij} + \delta\phi_{ij}$
end for
Find the current best
end while
Output the best results and pheromone distribution

Figure 7.1: Pseudo code of ant colony optimization.

$d_{ij} \propto 1/s_{ij}$, which implies that shorter routes will be selected due to their shorter traveling time, and thus the pheromone concentrations on these routes are higher. This is because the traveling time is shorter, and thus the less amount of the pheromone has been evaporated during this period.

This probability formula reflects the fact that ants would normally follow the paths with higher pheromone concentrations. In the simpler case when $\alpha = \beta = 1$, the probability of choosing a path by ants is proportional to the pheromone concentration on the path. The denominator normalizes the probability so that it is in the range between 0 and 1.

The pheromone concentration can change with time due to the evaporation of pheromone. Furthermore, the advantage of pheromone evaporation is that the system could avoid being trapped in local optima. If there is no evaporation, then the path randomly chosen by the first ants will become the preferred path as the attraction of other ants by their pheromone. For a constant rate γ of pheromone decay or evaporation, the pheromone concentration usually varies with time exponentially

$$\phi(t) = \phi_0 e^{-\gamma t}, \tag{7.2}$$

where ϕ_0 is the initial concentration of pheromone and t is time. If $\gamma t \ll 1$, then we have $\phi(t) \approx (1 - \gamma t)\phi_0$. For the unitary time increment $\Delta t = 1$, the evaporation can be approximated by $\phi^{t+1} \leftarrow (1-\gamma)\phi^t$. Therefore, we have the simplified pheromone update formula:

$$\phi_{ij}^{t+1} = (1-\gamma)\phi_{ij}^t + \delta\phi_{ij}^t, \tag{7.3}$$

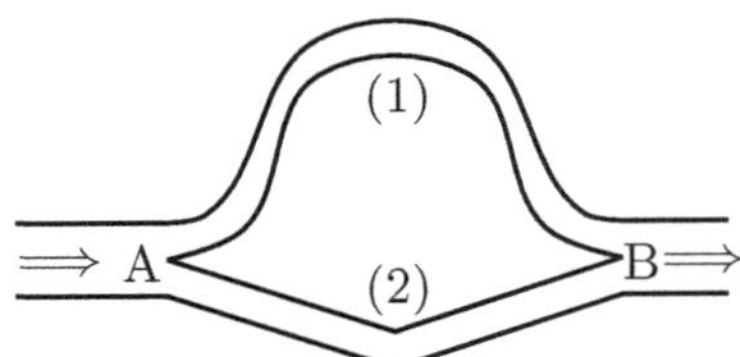

Figure 7.2: The double bridge problem for routing performance: route (2) is shorter than route (1).

where $\gamma \in [0, 1]$ is the rate of pheromone evaporation. The increment $\delta\phi_{ij}^t$ is the amount of pheromone deposited at time t along route i to j when an ant travels a distance L. Usually $\delta\phi_{ij}^t \propto 1/L$. If there are no ants on a route, then the pheromone deposit is zero.

There are other variations to this basic procedure. A possible acceleration scheme is to use some bounds of the pheromone concentration and only the ants with the current global best solution(s) are allowed to deposit pheromone. In addition, certain ranking of solution fitness can also be used. These are still topics of active research.

7.1.3 Double Bridge Problem

A standard test problem for ant colony optimization is the simplest double bridge problem with two branches (see Fig. 7.2) where route (2) is shorter than route (1). The angles of these two routes are equal at both point A and point B so that the ants have equal chance (or 50-50 probability) of choosing each route randomly at the initial stage at point A.

Initially, fifty percent of the ants would go along the longer route (1) and the pheromone evaporates at a constant rate, but the pheromone concentration will become smaller as route (1) is longer and thus takes more time to travel through. Conversely, the pheromone concentration on the shorter route will increase steadily. After some iterations, almost all the ants will move along the shorter route. Here we only use two routes at the node, it is straightforward to extend it to the multiple routes at a node. It is expected that only the shortest route will be chosen ultimately. As any complex network system is always made of individual nodes, this algorithms can be extended to solve complex routing problems reasonably efficiently. In fact, the ant colony algorithms have been successfully applied to the Internet routing problem, the traveling salesman problem, combinatorial optimization problems, and other NP-hard problems.

7.1.4 Virtual Ant Algorithm

As we know that ant colony optimization has successfully solved NP-hard problems such as the traveling salesman problem, it can also be extended to solve the standard optimization problems of multimodal functions. The only problem now is to figure out how the ants will move on an n-dimensional hyper-surface. For simplicity, we will discuss the 2-D case which can easily be extended to higher dimensions. On a 2D landscape, ants can move in any direction or $0° \sim 360°$, but this will cause some problems. How to update the pheromone at a particular point as there are infinite number of points. One solution is to track the history of each ant moves and record the locations consecutively, and the other approach is to use a moving neighborhood or window. The ants 'smell' the pheromone concentration of their neighborhood at any particular location.

In addition, we can limit the number of directions the ants can move by quantizing the directions. For example, ants are only allowed to move left and right, and up and down (only 4 directions). We will use this quantized approach here, which will make the implementation much simpler. Furthermore, the objective function or landscape can be encoded into virtual food so that ants will move to the best locations where the best food sources are. This will make the search process even more simpler. This simplified algorithm is called Virtual Ant Algorithm (VAA) developed by Xin-She Yang and his colleagues in 2006, which has been successfully applied to topological optimization problems in engineering.

It is worth pointing out that ant colony algorithms are a good tool for combinatorial and discrete optimization. They have the advantages over other stochastic algorithms such as genetic algorithms and simulated annealing in dealing with dynamical network routing problems.

7.2 BEE-INSPIRED ALGORITHMS

Bee algorithms form another class of algorithms which are closely related to the ant colony optimization. Bee algorithms are inspired by the foraging behaviour of honeybees. Several variants of bee algorithms have been formulated, including the honeybee algorithm (HBA), the virtual bee algorithm (VBA), the artificial bee colony (ABC) optimization, the honeybee-mating Algorithm (HBMA) and others.

7.2.1 Behavior of Honeybees

Honeybees live in a colony and they forage and store honey in their constructed colony. Honeybees can communicate by pheromone and 'waggle dance'. For example, an alarming bee may release a chemical message (pheromone) to stimulate attack response in other bees. Furthermore, when bees find a good food source and bring some nectar back to the hive, they will communicate the location of the food source by performing the so-called waggle dances as a signal system. Such signaling dances vary from species to species, however, they will try to recruit more bees by using directional dancing with varying strength so as to communicate the direction and distance of the found food resource.

For multiple food sources such as flower patches, studies show that a bee colony seems to be able to allocate forager bees among different flower patches so as to maximize their total nectar intake. In order to survive the winter, a bee colony typically has to collect and store extra nectar, about 15 to 50 kg. The efficiency of nectar collection is consequently very important from the evolution point of view. Experimental studies have also been carried out by researchers, including the important work by S Camazine, T Seeley and J Sney in early 1990s and lately by Quijano and K. Passino and their colleagues. Various algorithms can be designed if we learn from the natural behaviour of bee colonies.

7.2.2 Bee Algorithms

Over the last decade or so, nature-inspired bee algorithms have started to emerge as a promising and powerful tool. It is difficult to pinpoint the exact dates when the bee algorithms were first formulated. They were developed over a few years independently by several groups of researchers.

From the literature survey, it seems that the honey bee algorithm (HBA) was first formulated in around 2004 by Craig A Tovey at Georgia Tech in collaboration with Sunil Nakrani then at Oxford University to study a method to allocate computers among different clients and web-hosting servers. Later in 2004 and earlier 2005, Xin-She Yang at Cambridge University developed a virtual bee algorithm (VBA) to solve numerical optimization problems, and it can optimize both functions and discrete problems, though only functions with two parameters were given as examples. Slightly later in 2005, Haddad and Afshar and their colleagues presented a honeybee-mating optimization (HBMO) algorithm which was subsequently applied to reservoir modelling and clustering. Around the same time, D Karaboga in Turkey developed an artificial bee colony (ABC) algorithm for numerical function optimization and a comparison study was carried by in the same group later in 2006 and 2008. These bee algorithms have an increasing popularity.

The essence of bee algorithms are the communication or broadcasting ability of a bee to some neighbourhood bees so they can 'know' and follow a bee to the best source, locations or routes to complete the optimization task. The detailed implementation will depend on the actual algorithms, and they may differ slightly and vary with different variants. However, the essence of all the bee algorithms can be summarized as the pseudo code in Fig. 7.3.

7.2.3 Honeybee Algorithm

In the honeybee algorithm, forager bees are allocated to different food sources (or flower patches) so as to maximize the total nectar intake. The colony has to 'optimize' the overall efficiency of nectar collection, the allocation of the bees is thus depending on many factors such as the nectar richness and the proximity to the hive. This problem is similar to the allocation of web-hosting servers in the Internet which was in fact one of the first problems solved using bee-inspired algorithms by Nakrani and Tovey in 2004.

Let $w_i(j)$ be the strength of the waggle dance of bee i at time step $t = j$, the probability of an observer bee following the dancing bee to forage can be determined in many ways depending on the actual variant of algorithms. A simple way is given by

$$p_i = \frac{w_i^j}{\sum_{i=1}^{n_f} w_i^j}, \tag{7.4}$$

where n_f is the number of bees in foraging process. t is the pseudo time or foraging expedition. The number of observer bees is $N - n_f$ when N is the total number of bees. Alternatively, we can define an exploration probability of a Gaussian type $p_e = 1 - p_i = \exp[-w_i^2/2\sigma^2]$, where σ is the volatility of the bee colony, and it controls the exploration and diversity of the foraging sites. If there is no dancing (no food found), then $w_i \to 0$, and $p_e = 1$. So all the bee explore randomly.

In other variant algorithms when applying to discrete problems such as job scheduling, a forager bee will perform waggle dance with a duration $\tau = \gamma f_p$ where f_p is the profitability or the richness of the food site, and γ is a scaling factor. The profitability should be related to the objective function.

In addition, the rating of each route is ranked dynamically and the path with the highest number of bees become the preferred path. For a routing problem, the probability of choosing a route between any two nodes can take the form similar to the equation (7.1). That is

$$p_{ij} = \frac{w_{ij}^{\alpha} d_{ij}^{\beta}}{\sum_{i,j=1}^{n} w_{ij}^{\alpha} d_{ij}^{\beta}}, \tag{7.5}$$

where $\alpha > 0$ and $\beta > 0$ are the influence parameters. w_{ij} is the dance strength along route i to j, and d_{ij} is the desirability of the same route.

The honeybee algorithm, similar to the ant colony algorithms, is very efficient in dealing with discrete optimization problems such as routing and

Bee Algorithms

Objective function $f(\boldsymbol{x})$, $\boldsymbol{x} = (x_1, ..., x_n)^T$ *& constraints*
Encode $f(\boldsymbol{x})$ *into virtual nectar levels*
Define dance routine (strength, direction) or protocol
while *(criterion)*
  for *loop over all n dimensions*
    (or nodes for routing and scheduling problems)
  Generate new solutions
  Evaluate the new solutions
  end for
  Communicate and update the optimal solution set
end while
Decode and output the best results

Figure 7.3: Pseudo code of bee algorithms

scheduling. When dealing with continuous optimization problems, it is not straightforward, and some modifications are needed.

7.2.4 Virtual Bee Algorithm

The virtual bee algorithm (VBA), developed by Xin-She Yang in 2005, is an optimization algorithm specially formulated for solving both discrete and continuous problems. It has some similarity to the particle swarm optimization (PSO) to be discussed later, rather than a bee algorithm. In VBA, the continuous objective function is directly encoded as virtual nectar, and the solutions (or decision variables) are the locations of the nectar. The activities such as the waggle dance strength (or similar to the pheromone in the ant algorithms) are combined with the nectar concentration as the 'fitness' of the solutions. For a maximization problem, the objective function can be thought as virtual nectar, while for minimization, the nectar is formulated in such a way that the minimal value of the objective function corresponds to the highest nectar concentration.

For discrete problems, the objective function such as the shorter paths are encoded and linked with the profitability of the nectar explorations, which is in turn linked with the dance strength of forager bees. In this way, the virtual bee algorithm is similar to the honeybee algorithm. However, there is a fundamental difference from other bee algorithms. That is, VBA has a broadcasting ability of the current best. The current best location is 'known' to every bee so that this algorithm is more efficient. In this way, forager bees do not have to come back to the hive to tell other onlooker bees via 'waggle dance' so as to save time. Similar broadcasting ability is also used in the particle swarm optimization, especially in the accelerated PSO algorithms.

For a mixed type of problem when the decision variables can take both discrete and continuous values, the encoding of the objective function into nectar should be carefully implemented, so it can represent the objective effectively. This is still an active area of current research.

7.2.5 Artificial Bee Colony Optimization

The artificial bee colony (ABC) optimization algorithm was first developed by D. Karaboga in 2005. Since then, Karaboga and Basturk and their colleagues have systematically studied the performance of the ABC algorithm concerning unstrained optimization problems and its extension.

In the ABC algorithm, the bees in a colony are divided into three groups: employed bees (forager bees), onlooker bees (observer bees) and scouts. For each food source, there is only one employed bee. That is to say, the number of employed bees is equal to the number of food sources. The employed bee of an discarded food site is forced to become a scout for searching new food sources randomly. Employed bees share information with the onlooker bees in a hive so that onlooker bees can choose a food source to forage. Unlike the honey bee algorithm which has two groups of the bees (forager bees and observer bees), bees in ABC are more specialized.

For a given objective function $f(\boldsymbol{x})$, it can be encoded as $F(\boldsymbol{x})$ to represent the amount of nectar at location $\boldsymbol{x}$, thus the probability P_i of an onlooker bee choose to go to the preferred food source at $\boldsymbol{x}_i$ can be defined by $P_i = F(\boldsymbol{x}_i)/\sum_{j=1}^{S} F(\boldsymbol{x}_j)$ where S is the number of food sources. At a particular food source, the intake efficiency is determined by F/T where F is the amount of nectar and T is the time spent at the food source. If a food source is tried/foraged at a given number of explorations without improvement, then it is abandoned, and the bee at this location will move on randomly to explore new locations.

Various applications have been carried out recently in the last few years, including the combinatorial optimization, job scheduling, web-hosting allocation, and engineering design optimization. For continuous decision variables, their performance is still under active research. The handling of the pheromone requires substantial amount of memory and computing time. Is it possible to eliminate the pheromone and just use the roaming agents? The answer is yes. Particle swarm optimization is just the right kind of algorithm for such further modifications which will be discussed later in detail.

REFERENCES

1. A. Afshar, O. B. Haddad, M. A. Marino, B. J. Adams, Honey-bee mating optimization (HBMO) algorithm for optimal reservoir operation, *J. Franklin Institute*, **344**, 452-462 (2007).
2. B. Basturk and D. Karabogo, An artificial bee colony (ABC) algorithm for numerical function optimizaton, in: IEEE Swarm Intelligence Symposium 2006, May 12-14, Indianapolis, IN, USA, (2006).
3. E. Bonabeau, M. Dorigo, G. Theraulaz, *Swarm Intelligence: From Natural to Artificial Systems.* Oxford University Press, 1999.
4. M. Dorigo, *Optimization, Learning and Natural Algorithms*, PhD thesis, Politencnico di Milano, Italy, 1992.
5. M. Dorigo and T. Stützle, *Ant Colony Optimization*, MIT Press, 2004.
6. M. Fathian, B. Amiri, A. Maroosi, Application of honey-bee mating optimization algorithm on clustering, *Applied Mathematics and Computation*, **190**, 1502-1513 (2007).
7. D. Karaboga, An idea based on honey bee swarm for numerical optimization, Technical Report TR06, Erciyes University, Turkey, 2005.
8. D. Karaboga and B. Basturk, On the performance of artificial bee colony (ABC) algorithm, *Applied Soft Computing*,**8**, 687-697 (2008).
9. R. F. Moritz and E. Southwick, *Bees as superorganisms*, Springer,1992.
10. S. Nakrani and C. Tovey, On honey bees and dynamic server allocation in Internet hosting centers, *Adaptive Behaviour*, **12**, 223-240 (2004).
11. X. S. Yang, J. M. Lees, C. T. Morley, Application of virtual ant algorithms in the optimization of CFRP shear strengthened precracked structures, *Lecture Notes in Computer Sciences*, **3991**, 834-837 (2006).
12. X. S. Yang, Engineering optimization via nature-inspired virtual bee algorithms, IWINAC 2005, *Lecture Notes in Computer Science*, **3562**, 317-323 (2005).

Chapter 8

SWARM OPTIMIZATION

Particle swarm optimization (PSO) was developed by Kennedy and Eberhart in 1995, based on the swarm behaviour such as fish and bird schooling in nature. Since then, PSO has generated much wider interests, and forms an exciting, ever-expanding research subject, called swarm intelligence. PSO has been applied to almost every area in optimization, computational intelligence, and design/scheduling applications. There are at least two dozens of PSO variants, and hybrid algorithms by combining PSO with other existing algorithms are also increasingly popular.

8.1 SWARM INTELLIGENCE

Many algorithms such as ant colony algorithms and virtual ant algorithms use the behaviour of the so-called swarm intelligence. Particle swarm optimization may have some similarities with genetic algorithms and ant algorithms, but it is much simpler because it does not use mutation/crossover operators or pheromone. Instead, it uses the real-number randomness and the global communication among the swarm particles. In this sense, it is also easier to implement as there is no encoding or decoding of the parameters into binary strings as those in genetic algorithms (which can also use real-number strings).

This algorithm searches the space of an objective function by adjusting the trajectories of individual agents, called particles, as the piecewise paths formed by positional vectors in a quasi-stochastic manner. The movement of a swarming particle consists of two major components: a stochastic component and a deterministic component. Each particle is attracted toward the position of the current global best $\boldsymbol{g}^*$ and its own best location $\boldsymbol{x}_i^*$ in history, while at the same time it has a tendency to move randomly.

When a particle finds a location that is better than any previously found locations, then it updates it as the new current best for particle i. There is a current best for all n particles at any time t during iterations. The aim is to find the global best among all the current best solutions until the objective no longer improves or after a certain number of iterations. The movement of particles is schematically represented in Fig. 8.1 where $\boldsymbol{x}_i^*$ is

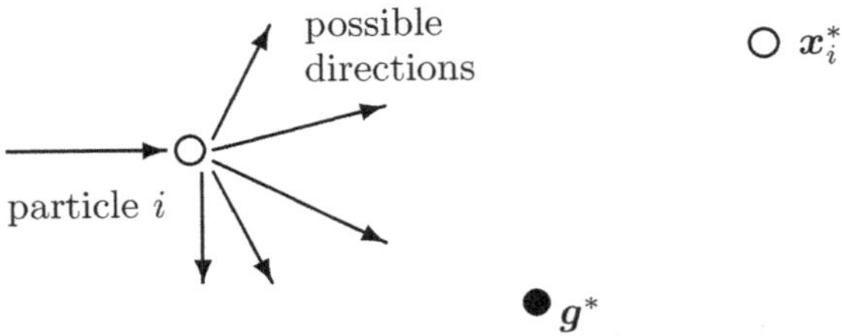

Figure 8.1: Schematic representation of the motion of a particle in PSO, moving towards the global best g^* and the current best x_i^* for each particle i.

Particle Swarm Optimization

Objective function $f(\boldsymbol{x})$, $\boldsymbol{x} = (x_1, ..., x_p)^T$
Initialize locations $\boldsymbol{x}_i$ *and velocity* $\boldsymbol{v}_i$ *of* n *particles.*
Find $\boldsymbol{g}^*$ *from* $\min\{f(\boldsymbol{x}_1), ..., f(\boldsymbol{x}_n)\}$ *(at* $t = 0$*)*
while *(criterion)*
  $t = t + 1$ *(pseudo time or iteration counter)*
  for *loop over all* n *particles and all* d *dimensions*
  Generate new velocity $\boldsymbol{v}_i^{t+1}$ *using equation (8.1)*
  Calculate new locations $\boldsymbol{x}_i^{t+1} = \boldsymbol{x}_i^t + \boldsymbol{v}_i^{t+1}$
  Evaluate objective functions at new locations $\boldsymbol{x}_i^{t+1}$
  Find the current best for each particle $\boldsymbol{x}_i^*$
  end for
  Fin the current global best $\boldsymbol{g}^*$
end while
Output the final results $\boldsymbol{x}_i^*$ *and* $\boldsymbol{g}^*$

Figure 8.2: Pseudo code of particle swarm optimization.

the current best for particle i, and $\boldsymbol{g}^* \approx \min\{f(\boldsymbol{x}_i)\}$ for $(i = 1, 2, ..., n)$ is the current global best.

8.2 PSO ALGORITHMS

The essential steps of the particle swarm optimization can be summarized as the pseudo code shown in Fig. 8.2.

Let $\boldsymbol{x}_i$ and $\boldsymbol{v}_i$ be the position vector and velocity for particle i, respectively. The new velocity vector is determined by the following formula

$$\boldsymbol{v}_i^{t+1} = \boldsymbol{v}_i^t + \alpha\epsilon_1 \odot [\boldsymbol{g}^* - \boldsymbol{x}_i^t] + \beta\epsilon_2 \odot [\boldsymbol{x}_i^* - \boldsymbol{x}_i^t]. \tag{8.1}$$

where ϵ_1 and ϵ_2 are two random vectors, and each entry taking the values between 0 and 1. The Hadamard product of two matrices $\boldsymbol{u} \odot \boldsymbol{v}$ is defined as the entrywise product, that is $[\boldsymbol{u} \odot \boldsymbol{v}]_{ij} = u_{ij}v_{ij}$. The parameters α and β are the learning parameters or acceleration constants, which can typically be taken as, say, $\alpha \approx \beta \approx 2$.

The initial locations of all particles should distribute relatively uniformly so that they can sample over most regions, which is especially important for multimodal problems. The initial velocity of a particle can be taken as zero, that is, $\boldsymbol{v}_i^{t=0} = 0$. The new position can then be updated by

$$\boldsymbol{x}_i^{t+1} = \boldsymbol{x}_i^t + \boldsymbol{v}_i^{t+1}. \tag{8.2}$$

Although $\boldsymbol{v}_i$ can be any values, it is usually bounded in some range $[0, \boldsymbol{v}_{\max}]$.

8.3 ACCELERATED PSO

There are many variants which extend the standard PSO algorithm, and the most noticeable improvement is probably to use inertia function $\theta(t)$ so that $\boldsymbol{v}_i^t$ is replaced by $\theta(t)\boldsymbol{v}_i^t$

$$\boldsymbol{v}_i^{t+1} = \theta \boldsymbol{v}_i^t + \alpha \epsilon_1 \odot [\boldsymbol{g}^* - \boldsymbol{x}_i^t] + \beta \epsilon_2 \odot [\boldsymbol{x}_i^* - \boldsymbol{x}_i^t], \tag{8.3}$$

where θ takes the values between 0 and 1. In the simplest case, the inertia function can be taken as a constant, typically $\theta \approx 0.5 \sim 0.9$. This is equivalent to introducing a virtual mass to stabilize the motion of the particles, and thus the algorithm is expected to converge more quickly.

The standard particle swarm optimization uses both the current global best $\boldsymbol{g}^*$ and the individual best $\boldsymbol{x}_i^*$. The reason of using the individual best is primarily to increase the diversity in the quality solutions, however, this diversity can be simulated using some randomness. Subsequently, there is no compelling reason for using the individual best, unless the optimization problem of interest is highly nonlinear and multimodal.

A simplified version which could accelerate the convergence of the algorithm is to use the global best only. Thus, in the accelerated particle swarm optimization, the velocity vector is generated by a simpler formula

$$\boldsymbol{v}_i^{t+1} = \boldsymbol{v}_i^t + \alpha(\epsilon - 1/2) + \beta(\boldsymbol{g}^* - \boldsymbol{x}_i^t), \tag{8.4}$$

where ϵ is a random variable with values from 0 to 1. Here the shift $1/2$ is purely out of convenience. We can also use a standard normal distribution $\alpha\epsilon_n$ where ϵ_n is drawn from $N(0,1)$ to replace the second term. The update of the position is simply

$$\boldsymbol{x}_i^{t+1} = \boldsymbol{x}_i^t + \boldsymbol{v}_i^{t+1}. \tag{8.5}$$

In order to increase the convergence even further, we can also write the update of the location in a single step

$$\boldsymbol{x}_i^{t+1} = (1-\beta)\boldsymbol{x}_i^t + \beta \boldsymbol{g}^* + \alpha \epsilon_n. \tag{8.6}$$

This simpler version will give the same order of convergence. The typical values for this accelerated PSO are $\alpha \approx 0.1 \sim 0.4$ and $\beta \approx 0.1 \sim 0.7$, though $\alpha \approx 0.2$ and $\beta \approx 0.5$ can be taken as the initial values for most unimodal objective functions. It is worth pointing out that the parameters α and β should in general be related to the scales of the independent variables $\boldsymbol{x}_i$ and the search domain.

A further improvement to the accelerated PSO is to reduce the randomness as iterations proceed. This means that we can use a monotonically decreasing function such as

$$\alpha = \alpha_0 e^{-\gamma t}, \tag{8.7}$$

or

$$\alpha = \alpha_0 \gamma^t, \qquad (0 < \gamma < 1), \tag{8.8}$$

where $\alpha_0 \approx 0.5 \sim 1$ is the initial value of the randomness parameter. Here t is the number of iterations or time steps. $0 < \gamma < 1$ is a control parameter. For example, in our implementation, we will use

$$\alpha = 0.7^t, \tag{8.9}$$

where $t \in [0, 10]$. Obviously, these parameters are fine-tuned to suit the current optimization problems as a demonstration.

8.4 IMPLEMENTATION

The accelerated particle swarm optimization has been implemented using both Matlab and Octave, and a simple program is provided below. This program can find the global optimal solution of most nonlinear functions in less a minute on most modern personal computers.

Now let us look at the 2D Michalewicz function

$$f(x, y) = -\{\sin(x)[\sin(\frac{x^2}{\pi})]^{2m} + \sin(y)[\sin(\frac{2y^2}{\pi})]^{2m}\},$$

where $m = 10$. The stationary conditions $f_x = f_y = 0$ require that

$$-\frac{4m}{\pi} x \sin(x) \cos(\frac{x^2}{\pi}) - \cos(x) \sin(\frac{x^2}{\pi}) = 0,$$

and

$$-\frac{8m}{\pi} y \sin(x) \cos(\frac{2y^2}{\pi}) - \cos(y) \sin(\frac{2y^2}{\pi}) = 0.$$

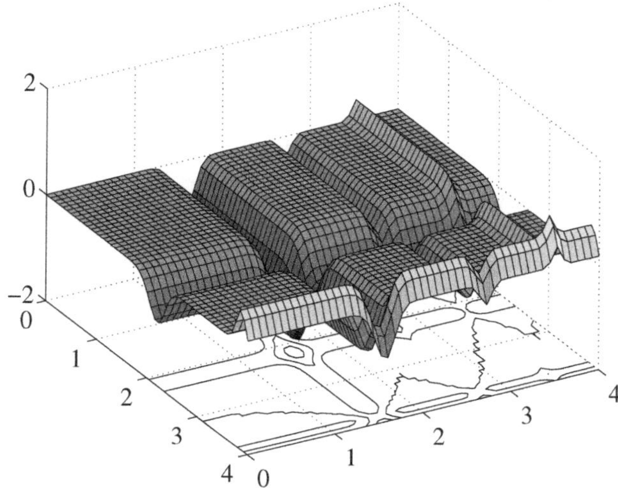

Figure 8.3: Michalewicz's function with a global minimum at about (2.20319, 1.57049).

The solution at (0, 0) is trivial, and the minimum $f^* \approx -1.801$ occurs at about (2.20319,1.57049) (see Fig. 8.3).

```
% The Particle Swarm Optimization
% (written by X S Yang, Cambridge University)
% Usage: pso(number_of_particles,Num_iterations)
%  eg:   best=pso_demo(20,10);
% where best=[xbest ybest zbest]  %an n by 3 matrix
%   xbest(i)/ybest(i) are the best at ith iteration

function [best]=pso_simpledemo(n,Num_iterations)
% n=number of particles
% Num_iterations=total number of iterations
if nargin<2,   Num_iterations=10;  end
if nargin<1,   n=20;              end
% Michalewicz Function f*=-1.801 at [2.20319,1.57049]
% Splitting two parts to avoid long lines in printing
str1='-sin(x)*(sin(x^2/3.14159))^20';
str2='-sin(y)*(sin(2*y^2/3.14159))^20';
funstr=strcat(str1,str2);
% Converting to an inline function and vectorization
f=vectorize(inline(funstr));
% range=[xmin xmax ymin ymax];
range=[0 4 0 4];
% ----------------------------------------------------
% Setting the parameters: alpha, beta
% Random amplitude of roaming particles alpha=[0,1]
```

```
% alpha=gamma^t=0.7^t;
% Speed of convergence (0->1)=(slow->fast)
beta=0.5;
% ----------------------------------------------------
% Grid values of the objective function
% These values are used for visualization only
Ngrid=100;
dx=(range(2)-range(1))/Ngrid;
dy=(range(4)-range(3))/Ngrid;
xgrid=range(1):dx:range(2); ygrid=range(3):dy:range(4);
[x,y]=meshgrid(xgrid,ygrid);
z=f(x,y);
% Display the shape of the function to be optimized
figure(1);
surfc(x,y,z);
% ----------------------------------------------------
best=zeros(Num_iterations,3);   % initialize history
% ----- Start Particle Swarm Optimization -----------
% generating the initial locations of n particles
[xn,yn]=init_pso(n,range);
% Display the paths of particles in a figure
% with a contour of the objective function
 figure(2);
% Start iterations
for i=1:Num_iterations,
% Show the contour of the function
  contour(x,y,z,15); hold on;
% Find the current best location (xo,yo)
zn=f(xn,yn);
zn_min=min(zn);
xo=min(xn(zn==zn_min));
yo=min(yn(zn==zn_min));
zo=min(zn(zn==zn_min));
% Trace the paths of all roaming particles
% Display these roaming particles
plot(xn,yn,'.',xo,yo,'*'); axis(range);
% The accelerated PSO with alpha=gamma^t
  gamma=0.7; alpha=gamma.^i;
% Move all the particles to new locations
[xn,yn]=pso_move(xn,yn,xo,yo,alpha,beta,range);
drawnow;
% Use "hold on" to display paths of particles
hold off;
% History
best(i,1)=xo; best(i,2)=yo; best(i,3)=zo;
end   %%%% end of iterations
% ----- All subfunctions are listed here -----
% Intial locations of n particles
```

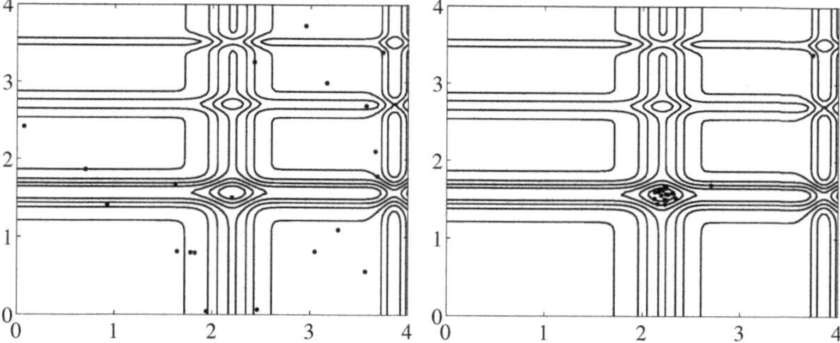

Figure 8.4: Initial and final locations of 20 particles after 10 iterations.

```
function [xn,yn]=init_pso(n,range)
xrange=range(2)-range(1); yrange=range(4)-range(3);
xn=rand(1,n)*xrange+range(1);
yn=rand(1,n)*yrange+range(3);
% Move all the particles toward (xo,yo)
function [xn,yn]=pso_move(xn,yn,xo,yo,a,b,range)
nn=size(yn,2);  %a=alpha, b=beta
xn=xn.*(1-b)+xo.*b+a.*(rand(1,nn)-0.5);
yn=yn.*(1-b)+yo.*b+a.*(rand(1,nn)-0.5);
[xn,yn]=findrange(xn,yn,range);
% Make sure the particles are within the range
function [xn,yn]=findrange(xn,yn,range)
nn=length(yn);
for i=1:nn,
   if xn(i)<=range(1), xn(i)=range(1); end
   if xn(i)>=range(2), xn(i)=range(2); end
   if yn(i)<=range(3), yn(i)=range(3); end
   if yn(i)>=range(4), yn(i)=range(4); end
end
```

If we run the program, we will get the global optimum after about 200 evaluations of the objective function (for 20 particles and 10 iterations). The results and the locations of the particles are shown in Fig. 8.4.

8.5 CONVERGENCE ANALYSIS

From the statistical point of view, each particle in PSO forms a Markov chain, though this Markov chain is biased towards to the current best, as the transition probability often leads to the acceptance of the move towards the current global best. In addition, the multiple Markov chains are interacting in terms of partly deterministic attraction movement. Therefore, the mathematical analysis concerning of the rate of convergence of PSO is very difficult, if not impossible. There is no doubt that any theoretical advance in understanding multiple interacting Markov chains will gain tremendous insightful in understanding how the PSO behaves and may consequently lead to the design of better or new PSO algorithms.

Mathematically, if we ignore the random factors, we can view the system formed by (8.1) and (8.2) as a dynamical system. If we focus on a single particle i and image there is only one particle in this system, then the global best $\boldsymbol{g}^*$ is the same as its current best $\boldsymbol{x}_i^*$. In this case, we have

$$\boldsymbol{v}_i^{t+1} = \boldsymbol{v}_i^t + \gamma(\boldsymbol{g}^* - \boldsymbol{x}_i^t), \qquad \gamma = \alpha + \beta, \tag{8.10}$$

and

$$\boldsymbol{x}_i^{t+1} = \boldsymbol{x}_i^t + \boldsymbol{v}_i^{t+1}. \tag{8.11}$$

Following an analysis of 1D dynamical system for particle swarm optimization by Clerc and Kennedy (2002), we can replace $\boldsymbol{g}^*$ by a parameter constant p so that we can see if or not the particle of interest will converge towards p. Now we can write the above system as a simple dynamical system

$$v(t+1) = v(t) + \gamma(p - x(t)), \qquad x(t+1) = x(t) + v(t+1). \tag{8.12}$$

For simplicity, we only focus on a single particle. By setting $u_t = p - x(t+1)$ and using the notations for dynamical systems, we have

$$v_{t+1} = v_t + \gamma u_t, \qquad u_{t+1} = -v_t + (1-\gamma)u_t, \tag{8.13}$$

or

$$Y_{t+1} = AY_t, \qquad A = \begin{pmatrix} 1 & \gamma \\ -1 & 1-\gamma \end{pmatrix}, \qquad Y_t = \begin{pmatrix} v_t \\ u_t \end{pmatrix}. \tag{8.14}$$

The general solution of this dynamical system can be written as

$$Y_t = Y_0 \exp[At]. \tag{8.15}$$

The main behaviour of this system can be characterized by the eigenvalues λ of A

$$\lambda_{1,2} = 1 - \frac{\gamma}{2} \pm \frac{\sqrt{\gamma^2 - 4\gamma}}{2}. \tag{8.16}$$

It can be seen clearly that $\gamma = 4$ leads to a bifurcation.

Following a straightforward analysis of this dynamical system, we can have three cases. For $0 < \gamma < 4$, cyclic and/or quasi-cyclic trajectories exist. In this case, when randomness is gradually reduced, some convergence can be observed. For $\gamma > 4$, non-cyclic behaviour can be expected and the distance from Y_t to the center $(0, 0)$ is monotonically increasing with t. In a special case $\gamma = 4$, some convergence behaviour can be observed.

For detailed analysis, please refer to Clerc and Kennedy (2002). Since p is linked with the global best, as the iterations continue, it can be expected that all particles will aggregate towards the the global best.

Various studies show that PSO algorithms can outperform genetic algorithms and other conventional algorithms for solving many optimization problems. This is partially due to that fact that the broadcasting ability of the current best estimates gives a better and quicker convergence towards the optimality. However, PSO algorithms are almost memoryless since they do not record the movement paths of each particle, and it is expected that it can be further improved using short-term memory in the similar fashion as that in Tabu search. Further development is under active research.

REFERENCES

1. A. Chatterjee and P. Siarry, Nonlinear inertia variation for dynamic adaptation in particle swarm optimization, *Comp. Oper. Research*, **33**, 859-871 (2006).
2. M. Clerc, J. Kennedy, The particle swarm - explosion, stability, and convergence in a multidimensional complex space, *IEEE Trans. Evolutionary Computation*, **6**, 58-73 (2002).
3. A. P. Engelbrecht, *Fundamentals of Computational Swarm Intelligence*, Wiley, 2005.
4. J. Kennedy and R. C. Eberhart, Particle swarm optimization, in: *Proc. of IEEE International Conference on Neural Networks*, Piscataway, NJ. pp. 1942-1948 (1995).
5. J. Kennedy, R. C. Eberhart, *Swarm intelligence*, Academic Press, 2001.
6. Swarm intelligence, http://www.swarmintelligence.org

Chapter 9

HARMONY SEARCH

9.1 HARMONICS AND FREQUENCIES

Harmony Search (HS) is a relatively new heuristic optimization algorithm and it was first developed by Z. W. Geem *et al.* in 2001. Since then, it has been applied to solve many optimization problems including function optimization, water distribution network, groundwater modelling, energy-saving dispatch, structural design, vehicle routing, and others. The possibility of combining harmony search with other algorithms such as particle swarm optimization and genetic algorithms has also been investigated.

Harmony search is a music-inspired metaheuristic optimization algorithm. It is inspired by the observation that the aim of music is to search for a perfect state of harmony. This harmony in music is analogous to find the optimality in an optimization process. The search process in optimization can be compared to a musician's improvisation process. This perfectly pleasing harmony is determined by the audio aesthetic standard.

The aesthetic quality of a musical instrument is essentially determined by its pitch (or frequency), timbre (or sound quality), and amplitude (or loudness). Timbre is largely determined by the harmonic content which is in turn determined by the waveforms or modulations of the sound signal. However, the harmonics it can generate will largely depend on the pitch or frequency range of the particular instrument.

Different notes have different frequencies. For example, the note A above middle C (or standard concert A4) has a fundamental frequency of $f_0 = 440$ Hz. As the speed of sound in dry air is about $v = 331 + 0.6T$ m/s where T is the temperature in degrees Celsius near 0°C. So at room temperature $T = 20$°C, the A4 note has a wavelength $\lambda = v/f_0 \approx 0.7795$ m. When we adjust the pitch, we are in fact trying to change the frequency. In music theory, pitch p in MIDI is often represented as a numerical scale (a linear pitch space) using the following formula

$$p = 69 + 12 \log_2(\frac{f}{440\text{Hz}}), \tag{9.1}$$

or

$$f = 440 \times 2^{(p-69)/12}, \tag{9.2}$$

Figure 9.1: Harmony of two notes with a frequency ratio of 2:3 and their waveform.

which means that the A4 notes has a pitch number 69. In this scale, octaves correspond to size 12 and semitone corresponds to size 1. Furthermore, the ratio of frequencies of two notes which are an octave apart is 2:1. Thus, the frequency of a note is doubled (halved) when it raised (lowered) by an octave. For example, A2 has a frequency of 110Hz, while A5 has a frequency of 880Hz.

The measurement of harmony when different pitches occurring simultaneously, like any aesthetic quality, is somewhat subjective. However, it is possible to use some standard estimation for harmony. The frequency ratio, pioneered by ancient Greek mathematician Pythagoras, is a good way for such estimations. For example, the octave with a ratio of 1:2 sounds pleasant when playing together, so are the notes with a ratio of 2:3 (see Fig. 9.1). However, it is unlikely for any random notes such as those shown in 9.2 to produce a pleasant harmony.

9.2 HARMONY SEARCH

Harmony search can be explained in more detail with the aid of the discussion of the improvisation process by a musician. When a musician is improvising, he or she has three possible choices: (1) play any famous piece of music (a series of pitches in harmony) exactly from his or her memory; (2) play something similar to a known piece (thus adjusting the pitch slightly); or (3) compose new or random notes. If we formalize these three options for optimization, we have three corresponding components: usage of harmony memory, pitch adjusting, and randomization.

The usage of harmony memory is important as it is similar to choose the best fit individuals in the genetic algorithms. This will ensure the best harmonies will be carried over to the new harmony memory. In order to use this memory more effectively, we can assign a parameter $r_{\text{accept}} \in [0, 1]$, called harmony memory accepting or considering rate. If this rate is too low, only few best harmonies are selected and it may converge too slowly.

Figure 9.2: Random music notes.

If this rate is extremely high (near 1), almost all the harmonies are used in the harmony memory, then other harmonies are not explored well, leading to potentially wrong solutions. Therefore, typically, $r_{\rm accept} = 0.7 \sim 0.95$.

To adjust the pitch slightly in the second component, we have to use a method such that it can adjust the frequency efficiently. In theory, the pitch can be adjusted linearly or nonlinearly, but in practice, linear adjustment is used. If $x_{\rm old}$ is the current solution (or pitch), then the new solution (pitch) $\boldsymbol{x}_{\rm new}$ is generated by

$$\boldsymbol{x}_{\rm new} = \boldsymbol{x}_{\rm old} + b_p \; (2 \; \texttt{rand} - 1), \tag{9.3}$$

where `rand` is a random number drawn from a uniform distribution $[0, 1]$. Here b_p is the bandwidth, which controls the local range of pitch adjustment. In fact, we can see that the pitch adjustment (9.3) is a random walk.

Pitch adjustment is similar to the mutation operator in genetic algorithms. We can assign a pitch-adjusting rate ($r_{\rm pa}$) to control the degree of the adjustment. If $r_{\rm pa}$ is too low, then there is rarely any change. If it is too high, then the algorithm may not converge at all. Thus, we usually use $r_{pa} = 0.1 \sim 0.5$ in most simulations.

The third component is the randomization, which is to increase the diversity of the solutions. Although adjusting pitch has a similar role, but it is limited to certain local pitch adjustment and thus corresponds to a local search. The use of randomization can drive the system further to explore various regions with high solution diversity so as to find the global optimality. So we have

$$p_a = p_{\rm lowerlimit} + p_{\rm range} * \texttt{rand}, \tag{9.4}$$

where $p_{\rm range} = p_{\rm upperlimit} - p_{\rm lowerlimit}$. Here `rand` is a random number generator in the range of 0 and 1.

The three components in harmony search can be summarized as the pseudo code shown in Fig. 9.3 where we can see that the probability of true randomization is

$$P_{\rm random} = 1 - r_{\rm accept}, \tag{9.5}$$

and the actual probability of pitch adjusting is

$$P_{\rm pitch} = r_{\rm accept} * r_{\rm pa}. \tag{9.6}$$

Harmony Search

Objective function $f(\boldsymbol{x})$, $\boldsymbol{x} = (x_1, ..., x_p)^T$
Generate initial harmonics (real number arrays)
Define pitch adjusting rate ($r_{\rm pa}$) *and pitch limits*
Define harmony memory accepting rate ($r_{\rm accept}$)
while *(t <Max number of iterations)*
    Generate new harmonics by accepting best harmonics
    Adjust pitch to get new harmonics (solutions)
    if *(rand>* $r_{\rm accept}$*),*
        choose an existing harmonic randomly
    else if *(rand>* $r_{\rm pa}$*),*
        adjust the pitch randomly within a bandwidth (9.3)
    else
        generate new harmonics via randomization (9.4)
    end if
    Accept the new harmonics (solutions) if better
end while
Find the current best estimates

Figure 9.3: Pseudo code of Harmony Search.

Furthermore, like genetic algorithms and particle swarm optimization, harmony search is not a gradient-based search, so it avoids most of the pitfalls of any gradient-based search algorithms. Thus, it has fewer mathematical requirements, and subsequently, it can be used to deal with complex objective functions whether continuous or discontinuous, linear or nonlinear, or stochastic with noise.

On the other hand, harmony search could be potentially more efficient than genetic algorithms because harmony search does not use binary encoding and decoding, but it does have multiple solution vectors. Therefore, HS can be faster during each iteration. The implementation of HS algorithm is also easier. In addition, there is evidence to suggest that HS is less sensitive to the chosen parameters, which means that we do not have to fine-tune these parameters to get quality solutions.

9.3 IMPLEMENTATION

Using the three components described in above section, we can implement the harmony search algorithm in Matlab/Octave.

For Rosenbrock's banana function

$$f(x, y) = (1 - x)^2 + 100(y - x^2)^2, \tag{9.7}$$

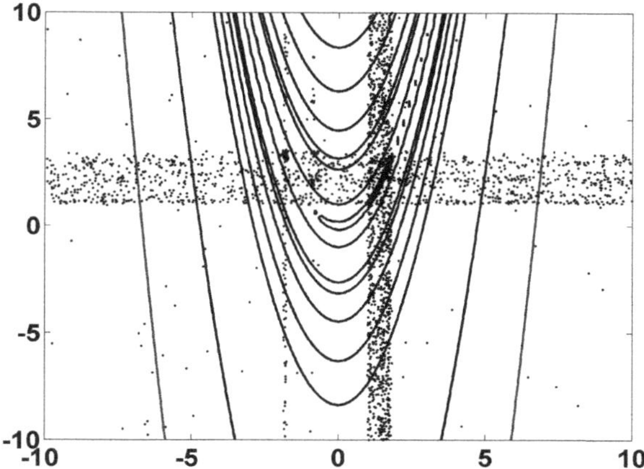

Figure 9.4: The variations of harmonies in harmony search.

within the domain

$$(x, y) \in [-10, 10] \times [-10, 10], \tag{9.8}$$

it has the global minimum $f_{\min} = 0$ at $(1, 1)$. The following Matlab/Octave program can be used to find its optimality.

```
% Harmony Search (Simple Demo) Matlab Program
% Written by X S Yang (Cambridge University)
% Usage: hs_simple
% or     hs_simple('x^2+(y-5)^2',25000);
function [solution,fbest]=hs_simple(funstr,MaxAttempt)
disp('It may take a few minutes ...');
% MaxAttempt=25000;  % Max number of Attempt
if nargin<2, MaxAttempt=25000; end
if nargin<1,
% Rosenbrock's Banana function with the
% global fmin=0 at (1,1).
funstr = '(1-x1)^2+100*(x2-x1^2)^2';
end
% Converting to an inline function
f=vectorize(inline(funstr));
ndim=2;  %Number of independent variables
% The range of the objective function
range(1,:)=[-10 10]; range(2,:)=[-10 10];
% Pitch range for pitch adjusting
pa_range=[200 200];
% Initial parameter setting
```

```
HS_size=20;          %Length of solution vector
HMacceptRate=0.95; %HM Accepting Rate
PArate=0.7;          %Pitch Adjusting rate
% Generating Initial Solution Vector
for i=1:HS_size,
   for j=1:ndim,
   x(j)=range(j,1)+(range(j,2)-range(j,1))*rand;
   end
   HM(i, :) = x;
   HMbest(i) = f(x(1), x(2));
end %% for i
% Starting the Harmony Search
for count = 1:MaxAttempt,
  for j = 1:ndim,
    if (rand >= HMacceptRate)
      % New Search via Randomization
      x(j)=range(j,1)+(range(j,2)-range(j,1))*rand;
    else
      % Harmony Memory Accepting Rate
      x(j) = HM(fix(HS_size*rand)+1,j);
      if (rand <= PArate)
      % Pitch Adjusting in a given range
      pa=(range(j,2)-range(j,1))/pa_range(j);
      x(j)= x(j)+pa*(rand-0.5);
      end
    end
  end %% for j
  % Evaluate the new solution
   fbest = f(x(1), x(2));
  % Find the best in the HS solution vector
   HSmaxNum = 1; HSminNum=1;
   HSmax = HMbest(1); HSmin=HMbest(1);
   for i = 2:HS_size,
      if HMbest(i) > HSmax,
        HSmaxNum = i;
        HSmax = HMbest(i);
      end
      if HMbest(i)<HSmin,
         HSminNum=i;
         HSmin=HMbest(i);
      end
   end
   % Updating the current solution if better
   if fbest < HSmax,
```

```
        HM(HSmaxNum, :) = x;
        HMbest(HSmaxNum) = fbest;
    end
    solution=x;  % Record the solution
end %% (end of harmony search)
```

The best estimate solution $(1.005, 1.0605)$ is obtained after 25000 iterations. On a modern desktop computer, it usually takes less than a minute. The variations of these solutions are shown in Fig. 9.4.

We have used r_{accept} =HMacceptRate= 0.95, and the pitch adjusting rate r_{pa} =PArate= 0.7. From Fig. 9.4, we can see that since the pitch adjustment is more intensive in local regions (two thin strips), it indeed indicates that the harmony search is more efficient than genetic algorithms. However, such comparison for different types of problem is still an area of active research.

Harmony search is emerging as a powerful algorithm, and its relevant literature is expanding. It is still an interesting area of active research.

REFERENCES

1. Z. W. Geem, J. H. Kim, and G. V. Loganathan, A new heuristic optimization algorithm: Harmony search, *Simulation*, **76**, 60-68 (2001).
2. Z. W. Geem, *Music-Inspired Harmony Search Algorithm: Theory and Applications*, Springer, (2009).
3. Z. W. Geem, *Recent Advances in Harmony Search Algorithm*, Studies in Computational Intelligence, Springer, 2010.
4. X. S. Yang, Harmony search as a metaheuristic algorithm, in: *Music-Inspired Harmony Search: Theory and Applications* (Eds Z. W. Geem), Springer, pp.1-14, 2009.

Chapter 10

FIREFLY ALGORITHM

10.1 BEHAVIOUR OF FIREFLIES

The flashing light of fireflies is an amazing sight in the summer sky in the tropical and temperate regions. There are about two thousand firefly species, and most fireflies produce short and rhythmic flashes. The pattern of flashes is often unique for a particular species. The flashing light is produced by a process of bioluminescence, and the true functions of such signaling systems are still being debated. However, two fundamental functions of such flashes are to attract mating partners (communication), and to attract potential prey. In addition, flashing may also serve as a protective warning mechanism to remind potential predators of the bitter taste of fireflies.

The rhythmic flash, the rate of flashing and the amount of time form part of the signal system that brings both sexes together. Females respond to a male's unique pattern of flashing in the same species, while in some species such as *Photuris*, female fireflies can eavesdrop on the bioluminescent courtship signals and even mimic the mating flashing pattern of other species so as to lure and eat the male fireflies who may mistake the flashes as a potential suitable mate. Some tropic fireflies can even synchronize their flashes, thus forming emerging biological self-organized behavior.

We know that the light intensity at a particular distance r from the light source obeys the inverse square law. That is to say, the light intensity I decreases as the distance r increases in terms of $I \propto 1/r^2$. Furthermore, the air absorbs light which becomes weaker and weaker as the distance increases. These two combined factors make most fireflies visual to a limit distance, usually several hundred meters at night, which is good enough for fireflies to communicate.

The flashing light can be formulated in such a way that it is associated with the objective function to be optimized, which makes it possible to formulate new optimization algorithms. In the rest of this chapter, we will first outline the basic formulation of the Firefly Algorithm (FA) and then discuss the implementation in detail.

Firefly Algorithm

Objective function $f(\boldsymbol{x})$, $\boldsymbol{x} = (x_1, ..., x_d)^T$
Generate initial population of fireflies $\boldsymbol{x}_i$ $(i = 1, 2, ..., n)$
Light intensity I_i *at* $\boldsymbol{x}_i$ *is determined by* $f(\boldsymbol{x}_i)$
Define light absorption coefficient γ
while *(t <MaxGeneration)*
for $i = 1 : n$ *all* n *fireflies*
  for $j = 1 : n$ *all* n *fireflies (inner loop)*
  if $(I_i < I_j)$, *Move firefly* i *towards* j; **end if**
    Vary attractiveness with distance r *via* $\exp[-\gamma r]$
    Evaluate new solutions and update light intensity
  end for j
end for i
Rank the fireflies and find the current global best $\boldsymbol{g}_*$
end while
Postprocess results and visualization

Figure 10.1: Pseudo code of the firefly algorithm (FA).

10.2 FIREFLY ALGORITHM

Now we can idealize some of the flashing characteristics of fireflies so as to develop firefly-inspired algorithms. For simplicity in describing our new Firefly Algorithm (FA) which was developed by Xin-She Yang at Cambridge University in 2007, we now use the following three idealized rules:

- All fireflies are unisex so that one firefly will be attracted to other fireflies regardless of their sex;
- Attractiveness is proportional to the their brightness, thus for any two flashing fireflies, the less brighter one will move towards the brighter one. The attractiveness is proportional to the brightness and they both decrease as their distance increases. If there is no brighter one than a particular firefly, it will move randomly;
- The brightness of a firefly is affected or determined by the landscape of the objective function.

For a maximization problem, the brightness can simply be proportional to the value of the objective function. Other forms of brightness can be defined in a similar way to the fitness function in genetic algorithms.

Based on these three rules, the basic steps of the firefly algorithm (FA) can be summarized as the pseudo code shown in Figure 11.1.

10.3 LIGHT INTENSITY AND ATTRACTIVENESS

In the firefly algorithm, there are two important issues: the variation of light intensity and formulation of the attractiveness. For simplicity, we can always assume that the attractiveness of a firefly is determined by its brightness which in turn is associated with the encoded objective function.

In the simplest case for maximum optimization problems, the brightness I of a firefly at a particular location $\boldsymbol{x}$ can be chosen as $I(\boldsymbol{x}) \propto f(\boldsymbol{x})$. However, the attractiveness β is relative, it should be seen in the eyes of the beholder or judged by the other fireflies. Thus, it will vary with the distance r_{ij} between firefly i and firefly j. In addition, light intensity decreases with the distance from its source, and light is also absorbed in the media, so we should allow the attractiveness to vary with the degree of absorption.

In the simplest form, the light intensity $I(r)$ varies according to the inverse square law

$$I(r) = \frac{I_s}{r^2}, \tag{10.1}$$

where I_s is the intensity at the source. For a given medium with a fixed light absorption coefficient γ, the light intensity I varies with the distance r. That is

$$I = I_0 e^{-\gamma r}, \tag{10.2}$$

where I_0 is the original light intensity. In order to avoid the singularity at $r = 0$ in the expression I_s/r^2, the combined effect of both the inverse square law and absorption can be approximated as the following Gaussian form

$$I(r) = I_0 e^{-\gamma r^2}. \tag{10.3}$$

As a firefly's attractiveness is proportional to the light intensity seen by adjacent fireflies, we can now define the attractiveness β of a firefly by

$$\beta = \beta_0 e^{-\gamma r^2}, \tag{10.4}$$

where β_0 is the attractiveness at $r = 0$. As it is often faster to calculate $1/(1 + r^2)$ than an exponential function, the above function, if necessary, can conveniently be approximated as

$$\beta = \frac{\beta_0}{1 + \gamma r^2}. \tag{10.5}$$

Both (10.4) and (10.5) define a characteristic distance $\Gamma = 1/\sqrt{\gamma}$ over which the attractiveness changes significantly from β_0 to $\beta_0 e^{-1}$ for equation (10.4) or $\beta_0/2$ for equation (10.5).

In the actual implementation, the attractiveness function $\beta(r)$ can be any monotonically decreasing functions such as the following generalized form

$$\beta(r) = \beta_0 e^{-\gamma r^m}, \qquad (m \geq 1). \tag{10.6}$$

For a fixed γ, the characteristic length becomes

$$\Gamma = \gamma^{-1/m} \to 1, \qquad m \to \infty. \tag{10.7}$$

Conversely, for a given length scale Γ in an optimization problem, the parameter γ can be used as a typical initial value. That is

$$\gamma = \frac{1}{\Gamma^m}. \tag{10.8}$$

The distance between any two fireflies i and j at $\boldsymbol{x}_i$ and $\boldsymbol{x}_j$, respectively, is the Cartesian distance

$$r_{ij} = \|\boldsymbol{x}_i - \boldsymbol{x}_j\| = \sqrt{\sum_{k=1}^{d} (x_{i,k} - x_{j,k})^2}, \tag{10.9}$$

where $x_{i,k}$ is the kth component of the spatial coordinate $\boldsymbol{x}_i$ of ith firefly. In 2-D case, we have

$$r_{ij} = \sqrt{(x_i - x_j)^2 + (y_i - y_j)^2}. \tag{10.10}$$

The movement of a firefly i is attracted to another more attractive (brighter) firefly j is determined by

$$\boldsymbol{x}_i = \boldsymbol{x}_i + \beta_0 e^{-\gamma r_{ij}^2} (\boldsymbol{x}_j - \boldsymbol{x}_i) + \alpha\, \boldsymbol{\epsilon}_i, \tag{10.11}$$

where the second term is due to the attraction. The third term is randomization with α being the randomization parameter, and $\boldsymbol{\epsilon}_i$ is a vector of random numbers drawn from a Gaussian distribution or uniform distribution. For example, the simplest form is $\boldsymbol{\epsilon}_i$ can be replaced by `rand` $- 1/2$ where `rand` is a random number generator uniformly distributed in $[0, 1]$. For most our implementation, we can take $\beta_0 = 1$ and $\alpha \in [0, 1]$.

It is worth pointing out that (10.11) is a random walk biased towards the brighter fireflies. If $\beta_0 = 0$, it becomes a simple random walk. Furthermore, the randomization term can easily be extended to other distributions such as Lévy flights.

The parameter γ now characterizes the variation of the attractiveness, and its value is crucially important in determining the speed of the convergence and how the FA algorithm behaves. In theory, $\gamma \in [0, \infty)$, but in practice, $\gamma = O(1)$ is determined by the characteristic length Γ of the system to be optimized. Thus, for most applications, it typically varies from 0.1 to 10.

10.4 SCALINGS AND ASYMPTOTICS

It is worth pointing out that the distance r defined above is *not* limited to the Euclidean distance. We can define other distance r in the n-dimensional hyperspace, depending on the type of problem of our interest. For example, for job scheduling problems, r can be defined as the time lag or time interval. For complicated networks such as the Internet and social networks, the distance r can be defined as the combination of the degree of local clustering and the average proximity of vertices. In fact, any measure that can effectively characterize the quantities of interest in the optimization problem can be used as the 'distance' r.

The typical scale Γ should be associated with the scale concerned in our optimization problem. If Γ is the typical scale for a given optimization problem, for a very large number of fireflies $n \gg k$ where k is the number of local optima, then the initial locations of these n fireflies should distribute relatively uniformly over the entire search space. As the iterations proceed, the fireflies would converge into all the local optima (including the global ones). By comparing the best solutions among all these optima, the global optima can easily be achieved. Our recent research suggests that it is possible to prove that the firefly algorithm will approach global optima when $n \to \infty$ and $t \gg 1$. In reality, it converges very quickly and this will be demonstrated later in this chapter.

There are two important limiting or asymptotic cases when $\gamma \to 0$ and $\gamma \to \infty$. For $\gamma \to 0$, the attractiveness is constant $\beta = \beta_0$ and $\Gamma \to \infty$, this is equivalent to saying that the light intensity does not decrease in an idealized sky. Thus, a flashing firefly can be seen anywhere in the domain. Thus, a single (usually global) optima can easily be reached. If we remove the inner loop for j in Figure 11.1 and replace $\boldsymbol{x}_j$ by the current global best $\boldsymbol{g}_*$, then the Firefly Algorithm becomes the special case of accelerated particle swarm optimization (PSO) discussed earlier. Subsequently, the efficiency of this special case is the same as that of PSO.

On the other hand, the limiting case $\gamma \to \infty$ leads to $\Gamma \to 0$ and $\beta(r) \to \delta(r)$ which is the Dirac delta function, which means that the attractiveness is almost zero in the sight of other fireflies. This is equivalent to the case where the fireflies roam in a very thick foggy region randomly. No other fireflies can be seen, and each firefly roams in a completely random way. Therefore, this corresponds to the completely random search method.

As the firefly algorithm is usually in the case between these two extremes, it is possible to adjust the parameter γ and α so that it can outperform both the random search and PSO. In fact, FA can find the global optima as well as the local optima simultaneously and effectively. This advantage will be demonstrated in detail later in the implementation.

A further advantage of FA is that different fireflies will work almost independently, it is thus particular suitable for parallel implementation. It is even better than genetic algorithms and PSO because fireflies aggregate more closely around each optimum. It can be expected that the interactions between different subregions are minimal in parallel implementation.

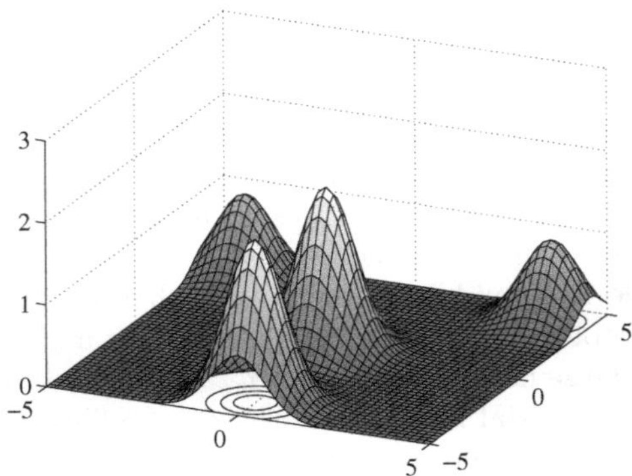

Figure 10.2: Landscape of a function with two equal global maxima.

10.5 IMPLEMENTATION

In order to demonstrate how the firefly algorithm works, we have implemented it in Matlab/Octave to be given later.

In order to show that both the global optima and local optima can be found simultaneously, we now use the following four-peak function

$$f(x,y) = e^{-(x-4)^2-(y-4)^2} + e^{-(x+4)^2-(y-4)^2} + 2[e^{-x^2-y^2} + e^{-x^2-(y+4)^2}],$$

where $(x, y) \in [-5, 5] \times [-5, 5]$. This function has four peaks. Two local peaks with $f = 1$ at $(-4, 4)$ and $(4, 4)$, and two global peaks with $f_{\max} = 2$ at $(0, 0)$ and $(0, -4)$, as shown in Figure 10.2. We can see that all these four optima can be found using 25 fireflies in about 20 generations (see Fig. 10.3). So the total number of function evaluations is about 500. This is much more efficient than most of existing metaheuristic algorithms.

```
% Firefly Algorithm by X S Yang (Cambridge University)
% Usage: ffa_demo([number_of_fireflies,MaxGeneration])
%  eg:   ffa_demo([12,50]);
function [best]=firefly_simple(instr)
% n=number of fireflies
% MaxGeneration=number of pseudo time steps
if nargin<1,    instr=[12 50];     end
n=instr(1);  MaxGeneration=instr(2);
rand('state',0);  % Reset the random generator
% ------ Four peak functions ---------------------
str1='exp(-(x-4)^2-(y-4)^2)+exp(-(x+4)^2-(y-4)^2)';
str2='+2*exp(-x^2-(y+4)^2)+2*exp(-x^2-y^2)';
funstr=strcat(str1,str2);
% Converting to an inline function
f=vectorize(inline(funstr));
```

```
% range=[xmin xmax ymin ymax];
range=[-5 5 -5 5];

% ------------------------------------------------
alpha=0.2;      % Randomness 0--1 (highly random)
gamma=1.0;      % Absorption coefficient
% ------------------------------------------------
% Grid values are used for display only
Ngrid=100;
dx=(range(2)-range(1))/Ngrid;
dy=(range(4)-range(3))/Ngrid;
[x,y]=meshgrid(range(1):dx:range(2),...
               range(3):dy:range(4));
z=f(x,y);
% Display the shape of the objective function
figure(1);    surfc(x,y,z);

% ------------------------------------------------
% generating the initial locations of n fireflies
[xn,yn,Lightn]=init_ffa(n,range);
% Display the paths of fireflies in a figure with
% contours of the function to be optimized
 figure(2);
% Iterations or pseudo time marching
for i=1:MaxGeneration,     %%%%% start iterations
% Show the contours of the function
 contour(x,y,z,15); hold on;
% Evaluate new solutions
zn=f(xn,yn);

% Ranking the fireflies by their light intensity
[Lightn,Index]=sort(zn);
xn=xn(Index); yn=yn(Index);
xo=xn;   yo=yn;    Lighto=Lightn;
% Trace the paths of all roaming  fireflies
plot(xn,yn,'.','markersize',10,'markerfacecolor','g');
% Move all fireflies to the better locations
[xn,yn]=ffa_move(xn,yn,Lightn,xo,yo,...
        Lighto,alpha,gamma,range);
drawnow;
% Use "hold on" to show the paths of fireflies
    hold off;
end   %%%%% end of iterations
best(:,1)=xo'; best(:,2)=yo'; best(:,3)=Lighto';

% ----- All subfunctions are listed here ---------
% The initial locations of n fireflies
function [xn,yn,Lightn]=init_ffa(n,range)
```

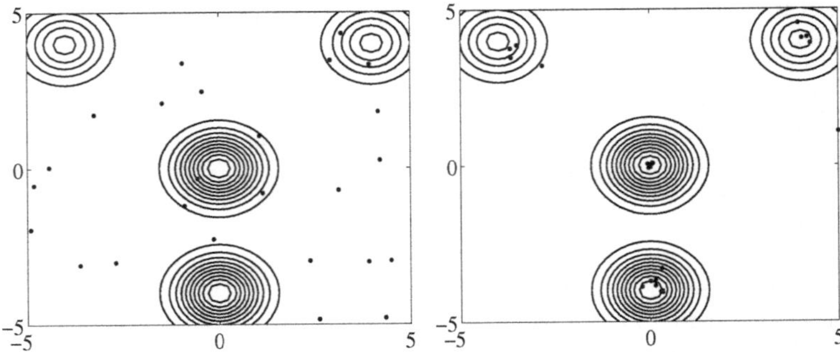

Figure 10.3: The initial locations of 25 fireflies (left) and their final locations after 20 iterations (right).

```
xrange=range(2)-range(1);
yrange=range(4)-range(3);
xn=rand(1,n)*xrange+range(1);
yn=rand(1,n)*yrange+range(3);
Lightn=zeros(size(yn));

% Move all fireflies toward brighter ones
function [xn,yn]=ffa_move(xn,yn,Lightn,xo,yo,...
    Lighto,alpha,gamma,range)
ni=size(yn,2); nj=size(yo,2);
for i=1:ni,
% The attractiveness parameter beta=exp(-gamma*r)
    for j=1:nj,
r=sqrt((xn(i)-xo(j))^2+(yn(i)-yo(j))^2);
if Lightn(i)<Lighto(j), % Brighter and more attractive
beta0=1;     beta=beta0*exp(-gamma*r.^2);
xn(i)=xn(i).*(1-beta)+xo(j).*beta+alpha.*(rand-0.5);
yn(i)=yn(i).*(1-beta)+yo(j).*beta+alpha.*(rand-0.5);
end
    end % end for j
end % end for i
[xn,yn]=findrange(xn,yn,range);

% Make sure the fireflies are within the range
function [xn,yn]=findrange(xn,yn,range)
for i=1:length(yn),
   if xn(i)<=range(1), xn(i)=range(1); end
   if xn(i)>=range(2), xn(i)=range(2); end
   if yn(i)<=range(3), yn(i)=range(3); end
   if yn(i)>=range(4), yn(i)=range(4); end
end
```

In the implementation, the values of the parameters are $\alpha = 0.2$, $\gamma = 1$ and $\beta_0 = 1$. Obviously, these parameters can be adjusted to suit for solving various problems with different scales.

10.6 FA VARIANTS

The basic firefly algorithm is very efficient, but we can see that the solutions are still changing as the optima are approaching. It is possible to improve the solution quality by reducing the randomness.

A further improvement on the convergence of the algorithm is to vary the randomization parameter α so that it decreases gradually as the optima are approaching. For example, we can use

$$\alpha = \alpha_\infty + (\alpha_0 - \alpha_\infty)e^{-t}, \tag{10.12}$$

where $t \in [0, t_{\max}]$ is the pseudo time for simulations and $t_{\max}$ is the maximum number of generations. α_0 is the initial randomization parameter while α_∞ is the final value. We can also use a similar function to the geometrical annealing schedule. That is

$$\alpha = \alpha_0 \theta^t, \tag{10.13}$$

where $\theta \in (0, 1]$ is the randomness reduction constant.

In addition, in the current version of the FA algorithm, we do not explicitly use the current global best $\boldsymbol{g}_*$, even though we only use it to decode the final best solutions. Our recent studies show that the efficiency may significantly improve if we add an extra term $\lambda \epsilon_i (x_i - \boldsymbol{g}_*)$ to the updating formula (10.11). Here λ is a parameter similar to α and β, and $\boldsymbol{\epsilon}_i$ is a vector of random numbers. These could form important topics for further research.

10.7 SPRING DESIGN

The design of a tension and compression spring is a well-known benchmark optimization problem. The main aim is to minimize the weight subject to constraints on deflection, stress, surge frequency and geometry. It involves three design variables: the wire diameter x_1, coil diameter x_2 and number/length of the coil x_3. This problem can be summarized as

$$\text{minimize } f(\boldsymbol{x}) = x_1^2 x_2 (2 + x_3), \tag{10.14}$$

subject to the following constraints

$$g_1(\boldsymbol{x}) = 1 - \frac{x_2^3 x_3}{71785 x_1^4} \leq 0,$$

$$g_2(\boldsymbol{x}) = \frac{4x_2^2 - x_1 x_2}{12566(x_1^3 x_2 - x_1^4)} + \frac{1}{5108 x_1^2} - 1 \leq 0,$$

$$g_3(\boldsymbol{x}) = 1 - \frac{140.45 x_1}{x_2^2 x_3} \leq 0,$$

$$g_4(\boldsymbol{x}) = \frac{x_1 + x_2}{1.5} - 1 \leq 0. \tag{10.15}$$

The simple bounds on the design variables are

$$0.05 \leq x_1 \leq 2.0, \qquad 0.25 \leq x_2 \leq 1.3, \qquad 2.0 \leq x_3 \leq 15.0. \tag{10.16}$$

The best solution found in the literature (e.g., Cagnina et al. 2008) is

$$\boldsymbol{x}_* = (0.051690, 0.356750, 11.287126), \tag{10.17}$$

with the objective

$$f(\boldsymbol{x}_*) = 0.012665. \tag{10.18}$$

We now provide the Matlab implementation of our firefly algorithm together with the penalty method for incorporating constraints. You may need a newer version of Matlab to deal with function handles. If you run the program a few times, you can get the above optimal solutions. It is even possible to produce better results if you experiment the program for a while.

```
% -------------------------------------------------------%
% Firefly Algorithm for constrained optimization         %
% by Xin-She Yang (Cambridge University) Copyright @2009 %
% -------------------------------------------------------%
function fa_mincon_demo

% parameters [n N_iteration alpha betamin gamma]
para=[40 150 0.5 0.2 1];

% This demo uses the Firefly Algorithm to solve the
% [Spring Design Problem as described by Cagnina et al.,
% Informatica, vol. 32, 319-326 (2008). ]

% Simple bounds/limits
disp('Solve the simple spring design problem ...');
Lb=[0.05 0.25 2.0];
Ub=[2.0 1.3 15.0];

% Initial random guess
u0=(Lb+Ub)/2;

[u,fval,NumEval]=ffa_mincon(@cost,@constraint,u0,Lb,Ub,para);
```

```
% Display results
bestsolution=u
bestojb=fval
total_number_of_function_evaluations=NumEval

%%% Put your own cost/objective function here --------%%%
%% Cost or Objective function
 function z=cost(x)
z=(2+x(3))*x(1)^2*x(2);

% Constrained optimization using penalty methods
% by changing f to F=f+ \sum lam_j*g^2_j*H_j(g_j)
% where H(g)=0 if g<=0 (true), =1 if g is false

%%% Put your own constraints here --------------------%%%
function [g,geq]=constraint(x)
% All nonlinear inequality constraints should be here
% If no inequality constraint at all, simple use g=[];
g(1)=1-x(2)^3*x(3)/(7178*x(1)^4);
tmpf=(4*x(2)^2-x(1)*x(2))/(12566*(x(2)*x(1)^3-x(1)^4));
g(2)=tmpf+1/(5108*x(1)^2)-1;
g(3)=1-140.45*x(1)/(x(2)^2*x(3));
g(4)=x(1)+x(2)-1.5;

% all nonlinear equality constraints should be here
% If no equality constraint at all, put geq=[] as follows
geq=[];

%%% End of the part to be modified -------------------%%%

%%% --------------------------------------------------%%%
%%% Do not modify the following codes unless you want %%%
%%% to improve its performance etc                    %%%
% -------------------------------------------------------
% ===Start of the Firefly Algorithm Implementation ======
% Inputs: fhandle => @cost (your own cost function,
%                    can be an external file  )
%     nonhandle => @constraint, all nonlinear constraints
%                    can be an external file or a function
%         Lb = lower bounds/limits
%         Ub = upper bounds/limits
%   para == optional (to control the Firefly algorithm)
% Outputs: nbest   = the best solution found so far
%          fbest   = the best objective value
%      NumEval = number of evaluations: n*MaxGeneration
% Optional:
% The alpha can be reduced (as to reduce the randomness)
% ---------------------------------------------------------
```

```
% Start FA
function [nbest,fbest,NumEval]...
          =ffa_mincon(fhandle,nonhandle,u0, Lb, Ub, para)
% Check input parameters (otherwise set as default values)
if nargin<6, para=[20 50 0.25 0.20 1]; end
if nargin<5, Ub=[]; end
if nargin<4, Lb=[]; end
if nargin<3,
disp('Usuage: FA_mincon(@cost, @constraint,u0,Lb,Ub,para)');
end

% n=number of fireflies
% MaxGeneration=number of pseudo time steps
% ------------------------------------------------
% alpha=0.25;      % Randomness 0--1 (highly random)
% betamn=0.20;     % minimum value of beta
% gamma=1;         % Absorption coefficient
% ------------------------------------------------
n=para(1);  MaxGeneration=para(2);
alpha=para(3); betamin=para(4); gamma=para(5);

% Total number of function evaluations
NumEval=n*MaxGeneration;

% Check if the upper bound & lower bound are the same size
if length(Lb) ~=length(Ub),
    disp('Simple bounds/limits are improper!');
    return
end

% Calcualte dimension
d=length(u0);

% Initial values of an array
zn=ones(n,1)*10^100;
% ------------------------------------------------
% generating the initial locations of n fireflies
[ns,Lightn]=init_ffa(n,d,Lb,Ub,u0);

% Iterations or pseudo time marching
for k=1:MaxGeneration,     %%%%% start iterations

% This line of reducing alpha is optional
 alpha=alpha_new(alpha,MaxGeneration);

% Evaluate new solutions (for all n fireflies)
for i=1:n,
```

```
   zn(i)=Fun(fhandle,nonhandle,ns(i,:));
   Lightn(i)=zn(i);
end

% Ranking fireflies by their light intensity/objectives
[Lightn,Index]=sort(zn);
ns_tmp=ns;
for i=1:n,
 ns(i,:)=ns_tmp(Index(i),:);
end

%% Find the current best
nso=ns; Lighto=Lightn;
nbest=ns(1,:); Lightbest=Lightn(1);

% For output only
fbest=Lightbest;

% Move all fireflies to the better locations
[ns]=ffa_move(n,d,ns,Lightn,nso,Lighto,nbest,...
      Lightbest,alpha,betamin,gamma,Lb,Ub);

end   %%%%% end of iterations

% -------------------------------------------------------
% ----- All the subfunctions are listed here ------------
% The initial locations of n fireflies
function [ns,Lightn]=init_ffa(n,d,Lb,Ub,u0)
  % if there are bounds/limits,
if length(Lb)>0,
   for i=1:n,
   ns(i,:)=Lb+(Ub-Lb).*rand(1,d);
   end
else
   % generate solutions around the random guess
   for i=1:n,
   ns(i,:)=u0+randn(1,d);
   end
end

% initial value before function evaluations
Lightn=ones(n,1)*10^100;

% Move all fireflies toward brighter ones
function [ns]=ffa_move(n,d,ns,Lightn,nso,Lighto,...
              nbest,Lightbest,alpha,betamin,gamma,Lb,Ub)
% Scaling of the system
scale=abs(Ub-Lb);
```

```
% Updating fireflies
for i=1:n,
% The attractiveness parameter beta=exp(-gamma*r)
   for j=1:n,
      r=sqrt(sum((ns(i,:)-ns(j,:)).^2));
      % Update moves
if Lightn(i)>Lighto(j), % Brighter and more attractive
   beta0=1; beta=(beta0-betamin)*exp(-gamma*r.^2)+betamin;
   tmf=alpha.*(rand(1,d)-0.5).*scale;
   ns(i,:)=ns(i,:).*(1-beta)+nso(j,:).*beta+tmpf;
      end
   end % end for j

end % end for i

% Check if the updated solutions/locations are within limits
[ns]=findlimits(n,ns,Lb,Ub);

% This function is optional, as it is not in the original FA
% The idea to reduce randomness is to increase the convergence,
% however, if you reduce randomness too quickly, then premature
% convergence can occur. So use with care.
function alpha=alpha_new(alpha,NGen)
% alpha_n=alpha_0(1-delta)^NGen=0.005
% alpha_0=0.9
delta=1-(0.005/0.9)^(1/NGen);
alpha=(1-delta)*alpha;

% Make sure the fireflies are within the bounds/limits
function [ns]=findlimits(n,ns,Lb,Ub)
for i=1:n,
     % Apply the lower bound
  ns_tmp=ns(i,:);
  I=ns_tmp<Lb;
  ns_tmp(I)=Lb(I);

  % Apply the upper bounds
  J=ns_tmp>Ub;
  ns_tmp(J)=Ub(J);
  % Update this new move
  ns(i,:)=ns_tmp;
end

% ---------------------------------------------
% d-dimensional objective function
function z=Fun(fhandle,nonhandle,u)
% Objective
```

```
z=fhandle(u);

% Apply nonlinear constraints by the penalty method
% Z=f+sum_k=1^N lam_k g_k^2 *H(g_k) where lam_k >> 1
z=z+getnonlinear(nonhandle,u);

function Z=getnonlinear(nonhandle,u)
Z=0;
% Penalty constant >> 1
lam=10^15; lameq=10^15;
% Get nonlinear constraints
[g,geq]=nonhandle(u);

% Apply inequality constraints as a penalty function
for k=1:length(g),
    Z=Z+ lam*g(k)^2*getH(g(k));
end
% Apply equality constraints (when geq=[], length->0)
for k=1:length(geq),
   Z=Z+lameq*geq(k)^2*geteqH(geq(k));
end

% Test if inequalities hold
% H(g) which is something like an index function
function H=getH(g)
if g<=0,
    H=0;
else
    H=1;
end

% Test if equalities hold
function H=geteqH(g)
if g==0,
    H=0;
else
    H=1;
end
%% ==== End of Firefly Algorithm implementation ======
```

REFERENCES

1. J. Arora, *Introduction to Optimum Design*, McGraw-Hill, (1989).

2. L. C. Cagnina, S. C. Esquivel, C. A. Coello, Solving engineering optimization problems with the simple constrained particle swarm optimizer, *Informatica*, **32**, 319-326 (2008).

3. S. Lukasik and S. Zak, Firefly algorithm for continuous constrained optimization tasks, ICCCI 2009, Lecture Notes in Artificial Intelligence (Eds. N. T. Ngugen et al.), **5796**, 97-106 (2009).

4. S. M. Lewis and C. K. Cratsley, Flash signal evolution, mate choice, and predation in fireflies, *Annual Review of Entomology*, **53**, 293-321 (2008).

5. C. O'Toole, *Firefly Encyclopedia of Insects and Spiders*, Firefly Books Ltd, 2002.

6. A. M. Reynolds and C. J. Rhodes, The Lévy flight paradigm: random search patterns and mechanisms, *Ecology*, **90**, 877-87 (2009).

7. E. G. Talbi, *Metaheuristics: From Design to Implementation*, Wiley, (2009).

8. X. S. Yang, *Nature-Inspired Metaheuristic Algorithms*, Luniver Press, (2008).

9. X. S. Yang, Firefly algorithms for multimodal optimization, in: *Stochastic Algorithms: Foundations and Applications*, SAGA 2009, Lecture Notes in Computer Science, **5792**, 169-178 (2009).

10. X. S. Yang, Firefly algorithm, Lévy flights and global optimization, in: *Research and Development in Intelligent Systems XXVI*, (Eds M. Bramer et al.), Springer, London, pp. 209-218 (2010).

Chapter 11

BAT ALGORITHM

Bat algorithm was formulated by Xin-She Yang in 2010, which showed some promising efficiency for global optimization. Interestingly, it can be viewed to link with harmony search and particle swarm optimization under appropriate conditions.

11.1 ECHOLOCATION OF BATS

11.1.1 Behaviour of microbats

Bats are fascinating animals. They are the only mammals with wings and they also have advanced capability of echolocation. It is estimated that there are about 1000 different species which account for up to 20% of all mammal species. Their size ranges from the tiny bumblebee bat (of about 1.5 to 2 g) to the giant bats with wingspan of about 2 m and weight up to about 1 kg. Microbats typically have forearm length of about 2.2 to 11 cm. Most bats uses echolocation to a certain degree; among all the species, microbats are a famous example as microbats use echolocation extensively, while megabats do not.

Most microbats are insectivores. Microbats use a type of sonar, called echolocation, to detect prey, avoid obstacles, and locate their roosting crevices in the dark. These bats emit a very loud sound pulse and listen for the echo that bounces back from the surrounding objects. Their pulses vary in properties and can be correlated with their hunting strategies, depending on the species. Most bats use short, frequency-modulated signals to sweep through about an octave, while others more often use constant-frequency signals for echolocation. Their signal bandwidth varies with species, and often increases by using more harmonics.

Studies show that microbats use the time delay from the emission and detection of the echo, the time difference between their two ears, and the loudness variations of the echoes to build up three dimensional scenario of the surrounding. They can detect the distance and orientation of the target, the type of prey, and even the moving speed of the prey such as small insects. Indeed, studies suggested that bats seem to be able to discriminate targets by the variations of the Doppler effect induced by the wing-flutter rates of the target insects.

11.1.2 Acoustics of Echolocation

Though each pulse only lasts a few thousandths of a second (up to about 8 to 10 ms), however, it has a constant frequency which is usually in the region of 25 kHz to 150 kHz. The typical range of frequencies for most bat species are in the region between 25 kHz and 100 kHz, though some species can emit higher frequencies up to 150 kHz. Each ultrasonic burst may last typically 5 to 20 ms, and microbats emit about 10 to 20 such sound bursts every second. When hunting for prey, the rate of pulse emission can be sped up to about 200 pulses per second when they fly near their prey. Such short sound bursts imply the fantastic ability of the signal processing power of bats. In fact, studies show the integration time of the bat ear is typically about 300 to 400 μs.

As the speed of sound in air is typically $v = 340$ m/s at room temperature, the wavelength λ of the ultrasonic sound bursts with a constant frequency f is given by

$$\lambda = \frac{v}{f}, \tag{11.1}$$

which is in the range of 2 mm to 14 mm for the typical frequency range from 25 kHz to 150 kHz. Such wavelengths are in the same order of their prey sizes.

Amazingly, the emitted pulse could be as loud as 110 dB, and, fortunately, they are in the ultrasonic region. The loudness also varies from the loudest when searching for prey and to a quieter base when homing towards the prey. The travelling range of such short pulses are typically a few metres, depending on the actual frequencies. Microbats can manage to avoid obstacles as small as thin human hairs.

Obviously, some bats have good eyesight, and most bats also have very sensitive smell sense. In reality, they will use all the senses as a combination to maximize the efficient detection of prey and smooth navigation. However, here we are only interested in the echolocation and the associated behaviour.

Such echolocation behaviour of microbats can be formulated in such a way that it can be associated with the objective function to be optimized, and this makes it possible to formulate new optimization algorithms. We will first outline the basic formulation of the Bat Algorithm (BA) and then discuss its implementation.

11.2 BAT ALGORITHM

If we idealize some of the echolocation characteristics of microbats, we can develop various bat-inspired algorithms or bat algorithms. For simplicity, we now use the following approximate or idealized rules:

1. All bats use echolocation to sense distance, and they also 'know' the difference between food/prey and background barriers;

2. Bats fly randomly with velocity $\boldsymbol{v}_i$ at position $\boldsymbol{x}_i$ with a fixed frequency $f_{\min}$ (or wavelength λ), varying wavelength λ (or frequency f) and loudness A_0 to search for prey. They can automatically adjust the wavelength (or frequency) of their emitted pulses and adjust the rate of pulse emission $r \in [0, 1]$, depending on the proximity of their target;

3. Although the loudness can vary in many ways, we assume that the loudness varies from a large (positive) A_0 to a minimum value $A_{\min}$.

Another obvious simplification is that no ray tracing is used in estimating the time delay and three dimensional topography. Though this might be a good feature for the application in computational geometry, however, we will not use this, as it is more computationally extensive in multidimensional cases.

In addition to these simplified assumptions, we also use the following approximations, for simplicity. In general the frequency f in a range $[f_{\min}, f_{\max}]$ corresponds to a range of wavelengths $[\lambda_{\min}, \lambda_{\max}]$. For example, a frequency range of [20 kHz, 500 kHz] corresponds to a range of wavelengths from 0.7 mm to 17 mm.

For a given problem, we can also use any wavelength for the ease of implementation. In the actual implementation, we can adjust the range by adjusting the frequencies (or wavelengths). The detectable range (or the largest wavelength) should be chosen such that it is comparable to the size of the domain of interest, and then toning down to smaller ranges. Furthermore, we do not necessarily have to use the wavelengths themselves at all, instead, we can also vary the frequency while fixing the wavelength λ. This is because λ and f are related, as λf is constant. We will use this later approach in our implementation.

For simplicity, we can assume $f \in [0, f_{\max}]$. We know that higher frequencies have short wavelengths and travel a shorter distance. For bats, the typical ranges are a few metres. The rate of pulse can simply be in the range of $[0, 1]$ where 0 means no pulses at all, and 1 means the maximum rate of pulse emission.

Based on the above approximations and idealization, the basic steps of the Bat Algorithm (BA) can be summarized as the pseudo code shown in Fig. 11.1.

11.2.1 Movement of Virtual Bats

In simulations, we have to use virtual bats. We have to define the rules how their positions $\boldsymbol{x}_i$ and velocities $\boldsymbol{v}_i$ in a d-dimensional search space are updated. The new solutions $\boldsymbol{x}_i^t$ and velocities $\boldsymbol{v}_i^t$ at time step t are given by

$$f_i = f_{\min} + (f_{\max} - f_{\min})\beta, \tag{11.2}$$

$$\boldsymbol{v}_i^t = \boldsymbol{v}_i^{t-1} + (\boldsymbol{x}_i^t - \boldsymbol{x}_*) f_i, \tag{11.3}$$

$$\boldsymbol{x}_i^t = \boldsymbol{x}_i^{t-1} + \boldsymbol{v}_i^t, \tag{11.4}$$

where $\beta \in [0, 1]$ is a random vector drawn from a uniform distribution. Here $\boldsymbol{x}_*$ is the current global best location (solution) which is located after comparing all the solutions among all the n bats. As the product $\lambda_i f_i$ is the velocity increment, we can use f_i (or λ_i) to adjust the velocity change while fixing the other factor λ_i (or f_i), depending on the type of the problem of interest. In our implementation, we will use $f_{\min} = 0$ and $f_{\max} = O(1)$, depending on the domain size of the problem of interest. Initially, each bat is randomly assigned a frequency which is drawn uniformly from $[f_{\min}, f_{\max}]$.

For the local search part, once a solution is selected among the current best solutions, a new solution for each bat is generated locally using random walk

$$\boldsymbol{x}_{\text{new}} = \boldsymbol{x}_{\text{old}} + \epsilon\, A^t, \tag{11.5}$$

Bat Algorithm

Initialize the bat population $\boldsymbol{x}_i$ $(i = 1, 2, ..., n)$ *and* $\boldsymbol{v}_i$
Initialize frequencies f_i, *pulse rates* r_i *and the loudness* A_i
while *(t <Max number of iterations)*
  Generate new solutions by adjusting frequency,
  and updating velocities and locations/solutions [(11.2) to (11.4)]
  if *(rand* $> r_i$*)*
    Select a solution among the best solutions
    Generate a local solution around the selected best solution
  end if
  Generate a new solution by flying randomly
  if *(rand* $< A_i$ *&* $f(\boldsymbol{x}_i) < f(\boldsymbol{x}_*)$*)*
    Accept the new solutions
    Increase r_i *and reduce* A_i
  end if
  Rank the bats and find the current best $\boldsymbol{x}_*$
end while

Figure 11.1: Pseudo code of the bat algorithm (BA).

where $\epsilon \in [-1, 1]$ is a random number, while $A^t = <A_i^t>$ is the average loudness of all the bats at this time step.

The update of the velocities and positions of bats have some similarity to the procedure in the standard particle swarm optimization, as f_i essentially controls the pace and range of the movement of the swarming particles. To a degree, BA can be considered as a balanced combination of the standard particle swarm optimization and the intensive local search controlled by the loudness and pulse rate.

11.2.2 Loudness and Pulse Emission

Furthermore, the loudness A_i and the rate r_i of pulse emission have to be updated accordingly as the iterations proceed. As the loudness usually decreases once a bat has found its prey, while the rate of pulse emission increases, the loudness can be chosen as any value of convenience. For example, we can use $A_0 = 100$ and $A_{\min} = 1$. For simplicity, we can also use $A_0 = 1$ and $A_{\min} = 0$, assuming $A_{\min} = 0$ means that a bat has just found the prey and temporarily stop emitting any sound. Now we have

$$A_i^{t+1} = \alpha A_i^t, \qquad r_i^{t+1} = r_i^0[1 - \exp(-\gamma t)], \tag{11.6}$$

where α and γ are constants. In fact, α is similar to the cooling factor of a cooling schedule in the simulated annealing discussed earlier in this book. For

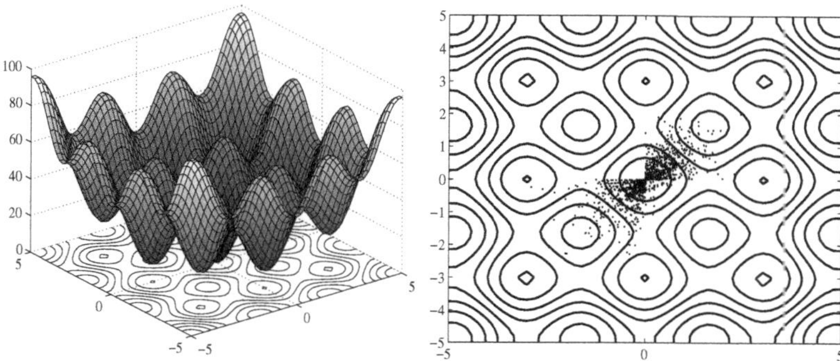

Figure 11.2: The eggcrate function (left) and the locations of 40 bats in the last ten iterations (right).

any $0 < \alpha < 1$ and $\gamma > 0$, we have

$$A_i^t \to 0, \quad r_i^t \to r_i^0, \quad \text{as } t \to \infty. \tag{11.7}$$

In the simplest case, we can use $\alpha = \gamma$, and we have used $\alpha = \gamma = 0.9$ in our simulations.

The choice of parameters requires some experimenting. Initially, each bat should have different values of loudness and pulse emission rate, and this can be achieved by randomization. For example, the initial loudness A_i^0 can typically be around $[1, 2]$, while the initial emission rate r_i^0 can be around zero, or any value $r_i^0 \in [0, 1]$ if using (11.6). Their loudness and emission rates will be updated only if the new solutions are improved, which means that these bats are moving towards the optimal solution.

11.3 VALIDATION AND DISCUSSIONS

From the pseudo code, it is relatively straightforward to implement the Bat Algorithm in any programming language. For the ease of visualization, we have implemented it using Matlab for various test functions.

There are many standard test functions for validating new algorithms. As a simple benchmark, let us look at the eggcrate function

$$f = x^2 + y^2 + 25(\sin^2 x + \sin^2 y), \quad (x, y) \in [-2\pi, 2\pi] \times [-2\pi, 2\pi].$$

We know that f has a global minimum $f_{\min} = 0$ at $(0, 0)$. In our implementation, we use $n = 25$ to 50 virtual bats, and $\alpha = 0.9$. For the multimodal eggcrate function, a snapshot of the last 10 iterations is shown in Fig. 11.2 where all bats move towards the global best $(0, 0)$.

The bat algorithm is much superior to other algorithms in terms of accuracy and efficiency (Yang and Deb, 2010). If we replace the variations of the frequency f_i by a random parameter and setting $A_i = 0$ and $r_i = 1$, the bat algorithm essentially becomes the standard particle swarm optimization (PSO).

Similarly, if we do not use the velocities, we use fixed loudness and rate: A_i and r_i. For example, $A_i = r_i = 0.7$, this algorithm is virtually reduced to a simple harmony search (HS), as the frequency/wavelength change is essentially the pitch adjustment, while the rate of pulse emission is similar to the harmonic acceptance rate (here with a twist) in the harmony search algorithm. The current studies imply that the proposed new algorithm is potentially more powerful and thus should be investigated further in many applications of engineering and industrial optimization problems.

11.4 IMPLEMENTATION

```
% -------------------------------------------------------%
% Bat-inspired algorithm for continuous optimization    %
% Programmed by Xin-She Yang @Cambridge University      %
% -------------------------------------------------------%
function [best,fmin,N_iter]=bat(para)
% Default parameters
if nargin<1,  para=[10 0.25 0.5];  end
n=para(1);        % Population size, typically 10 to 25
A=para(2);        % Loudness  (constant or decreasing)
r=para(3);        % Pulse rate (constant or decreasing)
% This frequency range determines the scalings
Qmin=0;           % Frequency minimum
Qmax=2;           % Frequency maximum
% Iteration parameters
tol=10^(-5);      % Stop tolerance
N_iter=0;         % Total number of function evaluations
% Dimension of the search variables
d=3;
% Initial arrays
Q=zeros(n,1);     % Frequency
v=zeros(n,d);     % Velocities
% Initialize the population/solutions
for i=1:n,
  Sol(i,:)=randn(1,d);
  Fitness(i)=Fun(Sol(i,:));
end
% Find the current best
[fmin,I]=min(Fitness);
best=Sol(I,:);
% Start the iterations -- Bat Algorithm
while (fmin>tol)
        % Loop over all bats/solutions
        for i=1:n,
          Q(i)=Qmin+(Qmin-Qmax)*rand;
          v(i,:)=v(i,:)+(Sol(i,:)-best)*Q(i);
          S(i,:)=Sol(i,:)+v(i,:);
```

```
          % Pulse rate
          if rand>r
              S(i,:)=best+0.01*randn(1,d);
          end

     % Evaluate new solutions
           Fnew=Fun(S(i,:));
     % If the solution improves or not too loudness
           if (Fnew<=Fitness(i)) & (rand<A) ,
                Sol(i,:)=S(i,:);
                Fitness(i)=Fnew;
           end

          % Update the current best
          if Fnew<=fmin,
                best=S(i,:);
                fmin=Fnew;
          end
        end
        N_iter=N_iter+n;
end
% Output/display
disp(['Number of evaluations: ',num2str(N_iter)]);
disp(['Best =',num2str(best),' fmin=',num2str(fmin)]);
% Objective function -- Rosenbrock's 3D function
function z=Fun(u)
z=(1-u(1))^2+100*(u(2)-u(1)^2)^2+(1-u(3))^2;
```

11.5 FURTHER TOPICS

From the formulation of the bat algorithm and its implementation and comparison, we can see that it is a very promising algorithm. It is potentially more powerful than particle swarm optimization and genetic algorithms as well as harmony search. The primary reason is that BA uses a good combination of major advantages of these algorithms in some way. Moreover, PSO and harmony search are the special cases of the bat algorithm under appropriate simplifications.

In addition, the fine adjustment of the parameters α and γ can affect the convergence rate of the bat algorithm. In fact, parameter α acts in a similar role as the cooling schedule in the simulated annealing. Though the implementation is slightly more complicated than those for many other metaheuristic algorithms; however, it does show that it utilizes a balanced combination of the advantages of existing successful algorithms with innovative feature based on the echolocation behaviour of bats. New solutions are generated by adjusting frequencies, loudness and pulse emission rates, while the proposed solution is accepted or not depends on the quality of the solutions controlled or characterized by loudness and pulse rate which are in turn related to the closeness or the fitness of the locations/solution to the global optimal solution.

The exciting results suggest that more studies will highly be needed to carry out the sensitivity analysis, to analyze the rate of algorithm convergence, and to improve the convergence rate even further. More extensive comparison studies with a more wide range of existing algorithms using much tough test functions in higher dimensions will pose more challenges to the algorithms, and thus such comparisons will potentially reveal the virtues and weakness of all the algorithms of interest.

An interesting extension will be to use different schemes of wavelength or frequency variations instead of the current linear implementation. In addition, the rates of pulse emission and loudness can also be varied in a more sophisticated manner. Another extension for discrete problems is to use the time delay between pulse emission and the echo bounced back. For example, in the travelling salesman problem, the distance between two adjacent nodes/cities can easily be coded as the time delay. As microbats use time difference between their two ears to obtain three-dimensional information, they can identify the type of prey and the velocity of a flying insect. Therefore, a further natural extension to the current bat algorithm would be to use the directional echolocation and Doppler effect, which may lead to even more interesting variants and new algorithms.

REFERENCES

1. Altringham, J. D., *Bats: Biology and Behaviour*, Oxford University Press, (1996).
2. Colin, T., *The Variety of Life*. Oxford University Press, (2000).
3. Geem, Z.W., Kim, J. H., Loganathan, G. V., A new heuristic optimization algorithm: Harmony search. *Simulation*, **76**, 60-68 (2001).
4. Goldberg, D., *Genetic Algorithms in Search, Optimization and Machine Learning*. Addison-Wesley, Reading, MA, (1989).
5. Richardson, P., *Bats*. Natural History Museum, London, (2008).
6. Richardson, P., The secrete life of bats. http://www.nhm.ac.uk
7. X. S. Yang, A new metaheuristic bat-inspired algorithm, in: *Nature Inspired Cooperative Strategies for Optimization* (NICSO 2010) (Eds. J. R. Gonzalez et al.), Springer, SCI **284**, 65-74 (2010).

Chapter 12

CUCKOO SEARCH

Cuckoo search (CS) is one of the latest nature-inspired metaheuristic algorithms, developed in 2009 by Xin-She Yang of Cambridge University and Suash Deb of C. V. Raman College of Engineering. CS is based on the brood parasitism of some cuckoo species. In addition, this algorithm is enhanced by the so-called Lévy flights, rather than by simple isotropic random walks. Recent studies show that CS is potentially far more efficient than PSO and genetic algorithms.

12.1 CUCKOO BREEDING BEHAVIOUR

Cuckoo are fascinating birds, not only because of the beautiful sounds they can make, but also because of their aggressive reproduction strategy. Some species such as the *ani* and *Guira* cuckoos lay their eggs in communal nests, though they may remove others' eggs to increase the hatching probability of their own eggs. Quite a number of species engage the obligate brood parasitism by laying their eggs in the nests of other host birds (often other species).

There are three basic types of brood parasitism: intraspecific brood parasitism, cooperative breeding, and nest takeover. Some host birds can engage direct conflict with the intruding cuckoos. If a host bird discovers the eggs are not their owns, they will either get rid of these alien eggs or simply abandon its nest and build a new nest elsewhere. Some cuckoo species such as the New World brood-parasitic *Tapera* have evolved in such a way that female parasitic cuckoos are often very specialized in the mimicry in colour and pattern of the eggs of a few chosen host species. This reduces the probability of their eggs being abandoned and thus increases their reproductivity.

In addition, the timing of egg-laying of some species is also amazing. Parasitic cuckoos often choose a nest where the host bird just laid its own eggs. In general, the cuckoo eggs hatch slightly earlier than their host eggs. Once the first cuckoo chick is hatched, the first instinct action it will take is to evict the host eggs by blindly propelling the eggs out of the nest, which increases the cuckoo chick's share of food provided by its host bird. Studies also show that a cuckoo chick can also mimic the call of host chicks to gain access to more feeding opportunity.

12.2 LÉVY FLIGHTS

On the other hand, various studies have shown that flight behaviour of many animals and insects has demonstrated the typical characteristics of Lévy flights. A recent study by Reynolds and Frye shows that fruit flies or *Drosophila melanogaster*, explore their landscape using a series of straight flight paths punctuated by a sudden 90° turn, leading to a Lévy-flight-style intermittent scale free search pattern. Studies on human behaviour such as the Ju/'hoansi hunter-gatherer foraging patterns also show the typical feature of Lévy flights. Even light can be related to Lévy flights. Subsequently, such behaviour has been applied to optimization and optimal search, and preliminary results show its promising capability.

12.3 CUCKOO SEARCH

For simplicity in describing our new Cuckoo Search, we now use the following three idealized rules:

- Each cuckoo lays one egg at a time, and dumps its egg in a randomly chosen nest;
- The best nests with high-quality eggs will be carried over to the next generations;
- The number of available host nests is fixed, and the egg laid by a cuckoo is discovered by the host bird with a probability $p_a \in [0, 1]$. In this case, the host bird can either get rid of the egg, or simply abandon the nest and build a completely new nest.

As a further approximation, this last assumption can be approximated by a fraction p_a of the n host nests are replaced by new nests (with new random solutions).

For a maximization problem, the quality or fitness of a solution can simply be proportional to the value of the objective function. Other forms of fitness can be defined in a similar way to the fitness function in genetic algorithms.

For the implementation point of view, we can use the following simple representations that each egg in a nest represents a solution, and each cuckoo can lay only one egg (thus representing one solution), the aim is to use the new and potentially better solutions (cuckoos) to replace a not-so-good solution in the nests. Obviously, this algorithm can be extended to the more complicated case where each nest has multiple eggs representing a set of solutions. For this present work, we will use the simplest approach where each nest has only a single egg. In this case, there is no distinction between egg, nest or cuckoo, as each nest corresponds to one egg which also represents one cuckoo.

Based on these three rules, the basic steps of the Cuckoo Search (CS) can be summarized as the pseudo code shown in Fig. 12.1.

When generating new solutions $\boldsymbol{x}^{(t+1)}$ for, say, a cuckoo i, a Lévy flight is performed

$$\boldsymbol{x}_i^{(t+1)} = \boldsymbol{x}_i^{(t)} + \alpha \oplus \text{Lévy}(\lambda), \tag{12.1}$$

where $\alpha > 0$ is the step size which should be related to the scales of the problem of interests. In most cases, we can use $\alpha = O(L/10)$ where L is the characteristic

Cuckoo Search via Lévy Flights

Objective function $f(\boldsymbol{x})$, $\boldsymbol{x} = (x_1, ..., x_d)^T$
Generate initial population of n *host nests* $\boldsymbol{x}_i$
while *(t <MaxGeneration) or (stop criterion)*
  Get a cuckoo randomly/generate a solution by Lévy flights
    and then evaluate its quality/fitness F_i
  Choose a nest among n *(say,* j*) randomly*
  if $(F_i > F_j)$,
    Replace j *by the new solution*
  end
  A fraction (p_a) *of worse nests are abandoned*
    and new ones/solutions are built/generated
  Keep best solutions (or nests with quality solutions)
  Rank the solutions and find the current best
end while
Postprocess results and visualization

Figure 12.1: Pseudo code of the Cuckoo Search (CS).

scale of the problem of interest. The above equation is essentially the stochastic equation for a random walk. In general, a random walk is a Markov chain whose next status/location only depends on the current location (the first term in the above equation) and the transition probability (the second term). The product $\oplus$ means entrywise multiplications. This entrywise product is similar to those used in PSO, but here the random walk via Lévy flight is more efficient in exploring the search space, as its step length is much longer in the long run.

The Lévy flight essentially provides a random walk whose random step length is drawn from a Lévy distribution

$$\text{Lévy} \sim u = t^{-\lambda}, \qquad (1 < \lambda \leq 3), \tag{12.2}$$

which has an infinite variance with an infinite mean. Here the steps essentially form a random walk process with a power-law step-length distribution with a heavy tail. Some of the new solutions should be generated by Lévy walk around the best solution obtained so far, this will speed up the local search. However, a substantial fraction of the new solutions should be generated by far field randomization and whose locations should be far enough from the current best solution, this will make sure that the system will not be trapped in a local optimum.

From a quick look, it seems that there is some similarity between CS and hill-climbing in combination with some large scale randomization. But there are some significant differences. Firstly, CS is a population-based algorithm, in a way similar to GA and PSO, but it uses some sort of elitism and/or selection similar to that used in harmony search. Secondly, the randomization in CS is more efficient as the step length is heavy-tailed, and any large step is possible. Thirdly, the number of parameters in CS to be tuned is fewer than GA and PSO, and thus it is potentially more generic to adapt to a wider class of optimization problems. In addition, each nest can represent a set of solutions, CS can thus be extended to the type of meta-population algorithms.

12.4 CHOICE OF PARAMETERS

After implementation, we have to validate the algorithm using test functions with analytical or known solutions. For example, one of the many test functions we have used is the bivariate Michalewicz function

$$f(x,y) = -\sin(x)\sin^{2m}(\frac{x^2}{\pi}) - \sin(y)\sin^{2m}(\frac{2y^2}{\pi}), \tag{12.3}$$

where $m = 10$ and $(x,y) \in [0,5] \times [0,5]$. This function has a global minimum $f_* \approx -1.8013$ at $(2.20319, 1.57049)$. This global optimum can easily be found using Cuckoo Search, and the results are shown in Fig. 12.2 where the final locations of the nests are also marked with $\diamond$ in the figure. Here we have used $n = 15$ nests, $\alpha = 1$ and $p_a = 0.25$. In most of our simulations, we have used $n = 15$ to 50.

From the figure, we can see that, as the optimum is approaching, most nests aggregate towards the global optimum. We also notice that the nests are also distributed at different (local) optima in the case of multimodal functions. This means that CS can find all the optima simultaneously if the number of nests are much higher than the number of local optima. This advantage may become more significant when dealing with multimodal and multiobjective optimization problems.

We have also tried to vary the number of host nests (or the population size n) and the probability p_a. We have used $n = 5, 10, 15,\ 20,\ 30,\ 40, 50,\ 100,$ $150,\ 250, 500$ and $p_a = 0,\ 0.01,\ 0.05, 0.1,\ 0.15, 0.2,\ 0.25,\ 0.3, 0.4, 0.5$. From our simulations, we found that $n = 15$ to 40 and $p_a = 0.25$ are sufficient for most optimization problems. Results and analysis also imply that the convergence rate, to some extent, is not sensitive to the parameters used. This means that the fine adjustment is not needed for any given problems.

12.5 IMPLEMENTATION

```
% -----------------------------------------------------
% Cuckoo algorithm by Xin-She Yang and Suasg Deb       %
% Programmed by Xin-She Yang at Cambridge University   %
% -----------------------------------------------------
function [bestsol,fval]=cuckoo_search(Ngen)
% Here Ngen is the max number of function evaluations
if nargin<1, Ngen=1500; end

% d-dimensions (any dimension)
d=2;
% Number of nests (or different solutions)
```

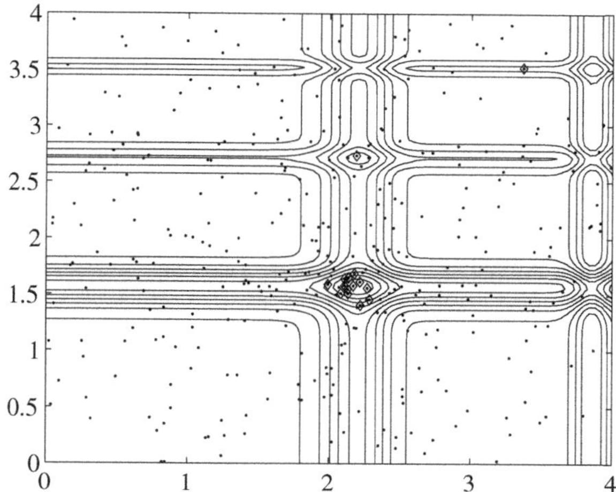

Figure 12.2: Search paths of nests using Cuckoo Search. The final locations of the nests are marked with $\diamond$ in the figure.

```
n=25;

% Discovery rate of alien eggs/solutions
pa=0.25;

% Random initial solutions
nest=randn(n,d);
fbest=ones(n,1)*10^(100);    % minimization problems
Kbest=1;

for j=1:Ngen,
    % Find the current best
    Kbest=get_best_nest(fbest);
    % Choose a random nest (avoid the current best)
    k=choose_a_nest(n,Kbest);
    bestnest=nest(Kbest,:)
    % Generate a new solution (but keep the current best)
    s=get_a_cuckoo(nest(k,:),bestnest);

    % Evaluate this solution
    fnew=fobj(s);
    if fnew<=fbest(k),
        fbest(k)=fnew;
        nest(k,:)=s;
    end
    % discovery and randomization
    if rand<pa,
```

```
        k=get_max_nest(fbest);
        s=emptyit(nest(k,:));
        nest(k,:)=s;
        fbest(k)=fobj(s);
    end
end

%% Post-optimization processing
%% Find the best and display
[fval,I]=min(fbest)
bestsol=nest(I,:)

%% Display all the nests
nest

%% --------- All subfunctions are listed below -----------
%% Choose a nest randomly
function k=choose_a_nest(n,Kbest)
k=floor(rand*n)+1;
% Avoid the best
if k==Kbest,
 k=mod(k+1,n)+1;
end

%% Get a cuckoo and generate new solutions by ramdom walk
function s=get_a_cuckoo(s,star)
% This is a random walk, which is less efficient
% than Levy flights. In addition, the step size
% should be a vector for problems with different scales.
% Here is the simplified implementation for demo only!
stepsize=0.05;
s=star+stepsize*randn(size(s));

%% Find the worse nest
function k=get_max_nest(fbest)
[fmax,k]=max(fbest);

%% Find the current best nest
function k=get_best_nest(fbest)
[fmin,k]=min(fbest);

%% Replace some (of the worst nests)
%% by constructing new solutions/nests
function s=emptyit(s)
% Again the step size should be varied
% Here is a simplified approach
s=s+0.05*randn(size(s));
```

```
% d-dimensional objective function
function z=fobj(u)
% Rosenbrock's function (in 2D)
% It has an optimal solution at (1.000,1.000)
z=(1-u(1))^2+100*(u(2)-u(1)^2)^2;
```

If we run this program using some standard test functions, we can observe that CS outperforms many existing algorithms such as GA and PSO. The primary reasons are: 1) a fine balance of randomization and intensification, and 2) fewer number of control parameters. As for any metaheuristic algorithms, a good balance of intensive local search and an efficient exploration of the whole search space will usually lead to a more efficient algorithm. On the other hand, there are only two parameters in this algorithm, the population size n, and p_a. Once n is fixed, p_a essentially controls the elitism and the balance of randomization and local search. Few parameters make an algorithm less complex and thus potentially more generic. Such observations deserve more systematic research and further elaboration in the future work.

It is worth pointing out that there are three ways to carry out randomization: uniform randomization, random walks and heavy-tailed walks. The simplest way is to use a uniform distribution so that new solutions are limited between upper and lower bounds. Random walks can be used for global randomization or local randomization, depending on the step size used in the implementation. Lévy flights are heavy-tailed, which is most suitable for the randomization on the global scale.

As an example for solving constrained optimization, we now solved the spring design problem discussed in the chapter on firefly algorithm. The Matlab code is given below

```
% Cuckoo Search for nonlinear constrained optimization
% Programmed by Xin-She Yang @ Cambridge University 2009
function [bestsol,fval]=cuckoo_spring(N_iter)
format long;
% number of iterations
if nargin<1, N_iter=15000; end
% Number of nests
n=25;
disp('Searching ... may take a minute or so ...');
% d variables and simple bounds
% Lower and upper bounds
Lb=[0.05 0.25 2.0];
Ub=[2.0  1.3  15.0];
% Number of variables
d=length(Lb);

% Discovery rate
pa=0.25;
% Random initial solutions
nest=init_cuckoo(n,d,Lb,Ub);
fbest=ones(n,1)*10^(10);   % minimization problems
```

```
Kbest=1;

% Start of the cuckoo search
for j=1:N_iter,
    % Find the best nest
    [fmin,Kbest]=get_best_nest(fbest);
    % Choose a nest randomly
    k=choose_a_nest(n,Kbest);
    bestnest=nest(Kbest,:) ;
    % Get a cuckoo with a new solution
    s=get_a_cuckoo(nest(k,:),bestnest,Lb,Ub);

    % Update if the solution improves
    fnew=fobj(s);
    if fnew<=fbest(k),
        fbest(k)=fnew;
        nest(k,:)=s;
    end

    % Discovery and randomization
    if rand<pa,
    k=get_max_nest(fbest);
    s=emptyit(nest(k,:),Lb,Ub);
    nest(k,:)=s;
    fbest(k)=fobj(s);
    end
end

%% Find the best
[fmin,I]=min(fbest)
bestsol=nest(I,:);

% Show all the nests
nest
% Display the best solution
bestsol, fmin

% Initial locations of all n cuckoos
function [guess]=init_cuckoo(n,d,Lb,Ub)
for i=1:n,
    guess(i,1:d)=Lb+rand(1,d).*(Ub-Lb);
end

%% Choose a nest randomly
function k=choose_a_nest(n,Kbest)
k=floor(rand*n)+1;
% Avoid the best
if k==Kbest,
```

```
 k=mod(k+1,n)+1;
end

%% Get a cuckoo with a new solution via a random walk
%% Note: Levy flights were not implemented in this demo
function s=get_a_cuckoo(s,star,Lb,Ub)
s=star+0.01*(Ub-Lb).*randn(size(s));
s=bounds(s,Lb,Ub);

%% Find the worse nest
function k=get_max_nest(fbest)
[fmax,k]=max(fbest);

%% Find the best nest
function [fmin,k]=get_best_nest(fbest)
[fmin,k]=min(fbest);

%% Replace an abandoned nest by constructing a new nest
function s=emptyit(s,Lb,Ub)
s=s+0.01*(Ub-Lb).*randn(size(s));
s=bounds(s,Lb,Ub);

% Check if bounds are met
function ns=bounds(ns,Lb,Ub)
% Apply the lower bound
  ns_tmp=ns;
  I=ns_tmp<Lb;
  ns_tmp(I)=Lb(I);
  % Apply the upper bounds
  J=ns_tmp>Ub;
  ns_tmp(J)=Ub(J);
% Update this new move
  ns=ns_tmp;

% d-dimensional objective function
function z=fobj(u)
% The well-known spring design problem
z=(2+u(3))*u(1)^2*u(2);
z=z+getnonlinear(u);

function Z=getnonlinear(u)
Z=0;
% Penalty constant
lam=10^15;

% Inequality constraints
g(1)=1-u(2)^3*u(3)/(71785*u(1)^4);
gtmp=(4*u(2)^2-u(1)*u(2))/(12566*(u(2)*u(1)^3-u(1)^4));
```

```
g(2)=gtmp+1/(5108*u(1)^2)-1;
g(3)=1-140.45*u(1)/(u(2)^2*u(3));
g(4)=(u(1)+u(2))/1.5-1;

% No equality constraint in this problem, so empty;
geq=[];

% Apply inequality constraints
for k=1:length(g),
    Z=Z+ lam*g(k)^2*getH(g(k));
end
% Apply equality constraints
for k=1:length(geq),
   Z=Z+lam*geq(k)^2*getHeq(geq(k));
end

% Test if inequalities hold
% Index function H(g) for inequalities
function H=getH(g)
if g<=0,
    H=0;
else
    H=1;
end
% Index function for equalities
function H=getHeq(geq)
if geq==0,
   H=0;
else
   H=1;
end
% ---------------- end ------------------------------
```

This potentially powerful optimization algorithm can easily be extended to study multiobjective optimization applications with various constraints, even to NP-hard problems. Further studies can focus on the sensitivity and parameter studies and their possible relationships with the convergence rate of the algorithm. Hybridization with other popular algorithms such as PSO and differential evolution will also be potentially fruitful.

REFERENCES

1. Barthelemy P., Bertolotti J., Wiersma D. S., A Lévy flight for light, *Nature*, **453**, 495-498 (2008).

2. Bradley D., Novel 'cuckoo search algorithm' beats particle swarm optimization in engineering design (news article), *Science Daily*, May 29, (2010). Also in *Scientific Computing* (magazine), 1 June 2010.

3. Brown C., Liebovitch L. S., Glendon R., Lévy flights in Dobe Ju/'hoansi foraging patterns, *Human Ecol.*, **35**, 129-138 (2007).

4. Chattopadhyay R., A study of test functions for optimization algorithms, *J. Opt. Theory Appl.*, **8**, 231-236 (1971).

5. Passino K. M., *Biomimicry of Bacterial Foraging for Distributed Optimization*, University Press, Princeton, New Jersey (2001).

6. Payne R. B., Sorenson M. D., and Klitz K., *The Cuckoos*, Oxford University Press, (2005).

7. Pavlyukevich I., Lévy flights, non-local search and simulated annealing, *J. Computational Physics*, **226**, 1830-1844 (2007).

8. Pavlyukevich I., Cooling down Lévy flights, *J. Phys. A:Math. Theor.*, **40**, 12299-12313 (2007).

9. Reynolds A. M. and Frye M. A., Free-flight odor tracking in Drosophila is consistent with an optimal intermittent scale-free search, *PLoS One*, **2**, e354 (2007).

10. A. M. Reynolds and C. J. Rhodes, The Lévy flight paradigm: random search patterns and mechanisms, *Ecology*, **90**, 877-87 (2009).

11. Schoen F., A wide class of test functions for global optimization, *J. Global Optimization*, **3**, 133-137, (1993).

12. Shlesinger M. F., Search research, *Nature*, **443**, 281-282 (2006).

13. Yang X. S. and Deb S., Cuckoo search via Lévy flights, in: *Proc. of World Congress on Nature & Biologically Inspired Computing* (NaBic 2009), IEEE Publications, USA, pp. 210-214 (2009).

14. Yang X. S. and Deb S,, Engineering optimization by cuckoo search, *Int. J. Math. Modelling & Numerical Optimisation*, **1**, 330-343 (2010).

Chapter 13

ANNS AND SUPPORT VECTOR MACHINE

As this book is an introduction to metaheuristic algorithms, our focus has always been on each algorithm. However, some algorithms cannot be classified as metaheuristic, but they are closely related to optimization. For this reason, we now briefly introduce artificial neural networks and support vector machine. As we will see, they are in essence optimization algorithms, working in different context.

13.1 ARTIFICIAL NEURAL NETWORKS

13.1.1 Artificial Neuron

The basic mathematical model of an artificial neuron was first proposed by W. McCulloch and W. Pitts in 1943, and this fundamental model is referred to as the McCulloch-Pitts model. Other models and neural networks are based on it.

An artificial neuron with n inputs or impulses and an output y_k will be activated if the signal strength reaches a certain threshold θ. Each input has a corresponding weight w_i (see Fig. 13.1). The output of this neuron is given by

$$y_l = \Phi\Big(\sum_{i=1}^{n} w_i u_i\Big), \tag{13.1}$$

where the weighted sum $\xi = \sum_{i=1}^{n} w_i u_i$ is the total signal strength, and Φ is the so-called activation function, which can be taken as a step function. That is, we have

$$\Phi(\xi) = \begin{cases} 1 & \text{if } \xi \geq \theta, \\ 0 & \text{if } \xi < \theta. \end{cases} \tag{13.2}$$

We can see that the output is only activated to a non-zero value if the overall signal strength is greater than the threshold θ.

The step function has discontinuity, sometimes, it is easier to use a nonlinear, smooth function, known as the Sigmoid function

$$S(\xi) = \frac{1}{1 + e^{-\xi}}, \tag{13.3}$$

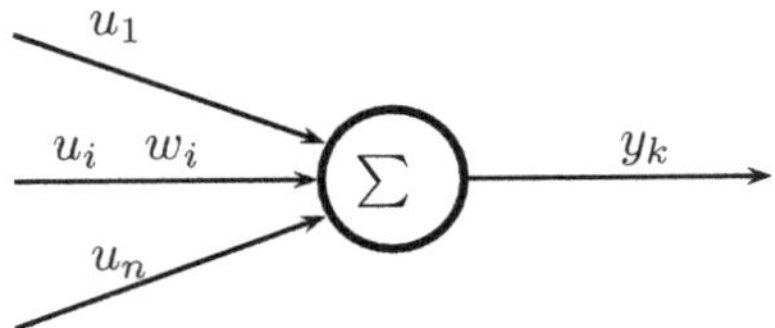

Figure 13.1: A simple neuron.

which approaches 1 as $U \to \infty$, and becomes 0 as $U \to -\infty$. An interesting property of this function is

$$S'(\xi) = S(\xi)[1 - S(\xi)]. \tag{13.4}$$

13.1.2 Neural Networks

A single neuron can only perform a simple task – on or off. Complex functions can be designed and performed using a network of interconnecting neurons or perceptrons. The structure of a network can be complicated, and one of the most widely used is to arrange them in a layered structure, with an input layer, an output layer, and one or more hidden layer (see Fig. 13.2). The connection strength between two neurons is represented by its corresponding weight. Some artificial neural networks (ANNs) can perform complex tasks, and can simulate complex mathematical models, even if there is no explicit functional form mathematically. Neural networks have developed over last few decades and have been applied in almost all areas of science and engineering.

The construction of a neural network involves the estimation of the suitable weights of a network system with some training/known data sets. The task of the training is to find the suitable weights ω_{ij} so that the neural networks not only can best-fit the known data, but also can predict outputs for new inputs. A good artificial neural network should be able to minimize both errors simultaneously – the fitting/learning errors and the prediction errors.

The errors can be defined as the difference between the calculated (or predicated) output o_k and real output y_k for all output neurons in the least-square sense

$$E = \frac{1}{2} \sum_{k=1}^{n_o} (o_k - y_k)^2. \tag{13.5}$$

Here the output o_k is a function of inputs/activations and weights. In order to minimize this error, we can use the standard minimization techniques to find the solutions of the weights.

A simple and yet efficient technique is the steepest descent method. For any initial random weights, the weight increment for w_{hk} is

$$\Delta w_{hk} = -\eta \frac{\partial E}{\partial w_{hk}} = -\eta \frac{\partial E}{\partial o_k} \frac{\partial o_k}{\partial w_{hk}}, \tag{13.6}$$

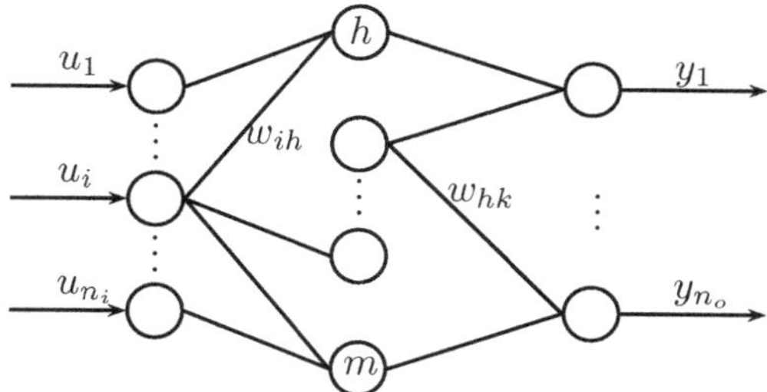

Figure 13.2: Schematic representation of a three-layer neural network with n_i inputs, m hidden nodes and n_o outputs.

where η is the learning rate. Typically, we can choose $\eta = 1$.

From

$$S_k = \sum_{h=1}^{m} w_{hk} o_h, \qquad (k = 1, 2, ..., n_o), \tag{13.7}$$

and

$$o_k = f(S_k) = \frac{1}{1 + e^{-S_k}}, \tag{13.8}$$

we have

$$f' = f(1 - f), \tag{13.9}$$

$$\frac{\partial o_k}{\partial w_{hk}} = \frac{\partial o_k}{\partial S_k} \frac{\partial S_k}{\partial w_{hk}} = o_k(1 - o_k) o_h, \tag{13.10}$$

and

$$\frac{\partial E}{\partial o_k} = (o_k - y_k). \tag{13.11}$$

Therefore, we have

$$\Delta w_{hk} = -\eta \delta_k o_h, \qquad \delta_k = o_k(1 - o_k)(o_k - y_k). \tag{13.12}$$

13.1.3 Back Propagation Algorithm

There are many ways of calculating weights by supervised learning. One of the simplest and widely used methods is to use the back propagation algorithm for training neural networks, often called back propagation neural networks (BPNNs).

The basic idea is to start from the output layer and propagate backwards so as to estimate and update the weights (see Fig. 13.3).

From any initial random weighting matrices w_{ih} (for connecting the input nodes to the hidden layer) and w_{hk} (for connecting the hidden layer to the output nodes), we can calculate the outputs of the hidden layer o_h

BPNN

Initialize weight matrices W_{ih} and W_{hk} randomly
for *all training data points*
  while *(residual errors are not zero)*
    Calculate the output for the hidden layer o_h using (13.13)
    Calculate the output for the output layer o_k using (13.14)
    Compute errors δ_k and δ_h using (13.15) and (13.16)
    Update weights w_{ih} and w_{hk} via (13.17) and (13.18)
  end while
end for

Figure 13.3: Pseudo code of back propagation neural networks.

$$o_h = \frac{1}{1+\exp[-\sum_{i=1}^{n_i} w_{ih} u_i]}, \qquad (h = 1, 2, ..., m), \tag{13.13}$$

and the outputs for the output nodes

$$o_k = \frac{1}{1+\exp[-\sum_{h=1}^{m} w_{hk} o_h]}, \qquad (k = 1, 2, ..., n_o). \tag{13.14}$$

The errors for the output nodes are given by

$$\delta_k = o_k(1 - o_k)(y_k - o_k), \qquad (k = 1, 2, ..., n_o), \tag{13.15}$$

where $y_k (k = 1, 2, ..., n_o)$ are the data (real outputs) for the inputs $u_i (i = 1, 2, ..., n_i)$. Similarly, the errors for the hidden nodes can be written as

$$\delta_h = o_h(1 - o_h) \sum_{k=1}^{n_o} w_{hk} \delta_k, \qquad (h = 1, 2, ..., m). \tag{13.16}$$

The updating formulae for weights at iteration t are

$$w_{hk}^{t+1} = w_{hk}^{t} + \eta \delta_k o_h, \tag{13.17}$$

and

$$w_{ih}^{t+1} = w_{ih}^{t} + \eta \delta_h u_i, \tag{13.18}$$

where $0 < \eta \leq 1$ is the learning rate.

Here we can see that the weight increments are

$$\Delta w_{ih} = \eta \delta_h u_i, \tag{13.19}$$

with similar updating formulae for w_{hk}. An improved version is to use the so-called weight momentum α to increase the learning efficiency

$$\Delta w_{ih} = \eta \delta_h u_i + \alpha w_{ih}(\tau - 1), \tag{13.20}$$

where τ is an extra parameter.

There are many good software packages for artificial neural networks, and there are dozens of good books fully dedicated to implementation. Therefore, we will not provide any code here.

13.2 SUPPORT VECTOR MACHINE

Support vector machine is a powerful tool which becomes increasingly popular in classifications, data mining, pattern recognition, artificial intelligence, and optimization.

13.2.1 Classifications

In many applications, the aim is to separate some complex data into different categories. For example, in pattern recognition, we may need to simply separate circles from squares. That is to label them into two different classes. In other applications, we have to answer a yes-no question, which is a binary classification.

If there are n different classes, we can in principle first classify them into two classes: class, say k, and non-class k. We then focus on the non-class k and divide them into two different classes, and so on and so forth.

Mathematically speaking, for a given (but scattered) data set, the objective is to separate them into different regions/domains or types. In the simplest case, the outputs are just class either A or B; that is, either +1 or −1.

13.2.2 Statistical Learning Theory

For the case of two-class classifications, we have the learning examples or data as $(\boldsymbol{x}_i, y_i)$ where $i = 1, 2, ..., n$ and $y_i \in \{-1, +1\}$. The aim of the learning is to find a function $f_\alpha(\boldsymbol{x})$ from allowable functions $\{f_\alpha : \alpha \in \Omega\}$ such that

$$f_\alpha(\boldsymbol{x}_i) \mapsto y_i, \qquad (i = 1, 2, ..., n), \tag{13.21}$$

and that the expected risk $E(\alpha)$ is minimal. That is the minimization of the risk

$$E(\alpha) = \frac{1}{2} \int |f_\alpha(x) - y| dP(\boldsymbol{x}, y), \tag{13.22}$$

where $P(\boldsymbol{x}, y)$ is an unknown probability distribution, which makes it impossible to calculate $E(\alpha)$ directly. A simple approach is to use the so-called empirical risk

$$E_p(\alpha) \approx \frac{1}{n} \sum_{i=1}^{n} \frac{1}{2} \Big| f_\alpha(\boldsymbol{x}_i) - y_i \Big|. \tag{13.23}$$

A main drawback of this approach is that a small risk or error on the training set does not necessarily guarantee a small error on prediction if the number n of training data is small.

In the framework of structural risk minimization and statistical learning theory, there exists an upper bound for such errors. For a given probability of at

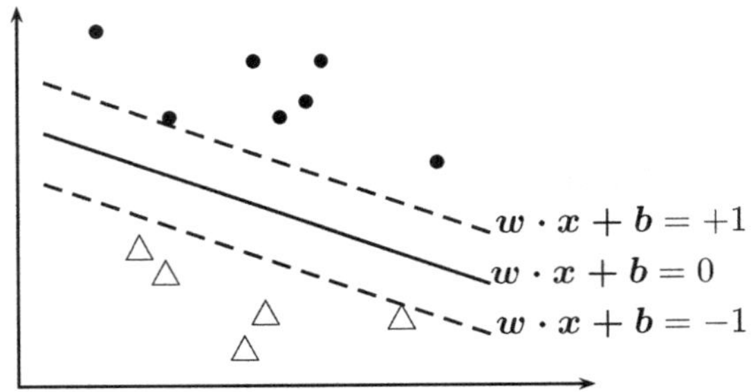

Figure 13.4: Hyperplane, maximum margins and linear support vector machine (SVM).

least $1 - p$, the Vapnik bound for the errors can be written as

$$E(\alpha) \leq E_p(\alpha) + \phi(\frac{h}{n}, \frac{\log(p)}{n}), \tag{13.24}$$

where

$$\phi(\frac{h}{n}, \frac{\log(p)}{n}) = \sqrt{\frac{1}{n}\Big[h(\log \frac{2n}{h} + 1) - \log(\frac{p}{4})\Big]}. \tag{13.25}$$

Here h is a parameter, often referred to as the Vapnik-Chervonenskis dimension (or simply VC-dimension). This dimension describes the capacity for prediction of the function set f_α. In the simplest binary classification with only two values of $+1$ and -1, h is essentially the maximum number of points which can be classified into two distinct classes in all possible 2^h combinations.

13.2.3 Linear Support Vector Machine

The basic idea of classification is to try to separate different samples into different classes. For binary classifications such as the triangles and spheres (or solid dots) as shown in Fig. 13.4, we intend to construct a hyperplane

$$\boldsymbol{w} \cdot \boldsymbol{x} + \boldsymbol{b} = 0, \tag{13.26}$$

so that these samples can be divided into classes with triangles on one side and the spheres on the other side. Here the normal vector $\boldsymbol{w}$ and $\boldsymbol{b}$ have the same size as $\boldsymbol{x}$, and they can be determined using the data, though the method of determining them is not straightforward. This requires the existence of a hyperplane; otherwise, this approach will not work. In this case, we have to use other methods.

In essence, if we can construct such a hyperplane, we should construct two hyperplanes (shown as dashed lines) so that the two hyperplanes should be as far away as possible and no samples should be between these two planes. Mathematically, this is equivalent to two equations

$$\boldsymbol{w} \cdot \boldsymbol{x} + \boldsymbol{b} = +1, \tag{13.27}$$

and

$$\boldsymbol{w} \cdot \boldsymbol{x} + \boldsymbol{b} = -1. \tag{13.28}$$

From these two equations, it is straightforward to verify that the normal (perpendicular) distance between these two hyperplanes is related to the norm $||\boldsymbol{w}||$ via

$$d = \frac{2}{||\boldsymbol{w}||}. \tag{13.29}$$

A main objective of constructing these two hyperplanes is to maximize the distance or the margin between the two planes. The maximization of d is equivalent to the minimization of $||w||$ or more conveniently $||w||^2$. From the optimization point of view, the maximization of margins can be written as

$$\text{minimize } \frac{1}{2}||\boldsymbol{w}||^2 = \frac{1}{2}(\boldsymbol{w} \cdot \boldsymbol{w}). \tag{13.30}$$

If we can classify all the samples completely, for any sample $(\boldsymbol{x}_i, y_i)$ where $i = 1, 2, ..., n$, we have

$$\boldsymbol{w} \cdot \boldsymbol{x}_i + \boldsymbol{b} \geq +1, \qquad \text{if } (\boldsymbol{x}_i, y_i) \in \text{one class}, \tag{13.31}$$

and

$$\boldsymbol{w} \cdot \boldsymbol{x}_i + \boldsymbol{b} \leq -1, \qquad \text{if } (\boldsymbol{x}_i, y_i) \in \text{the other class}. \tag{13.32}$$

As $y_i \in \{+1, -1\}$, the above two equations can be combined as

$$y_i(\boldsymbol{w} \cdot \boldsymbol{x}_i + \boldsymbol{b}) \geq 1, \qquad (i = 1, 2, ..., n). \tag{13.33}$$

However, in reality, it is not always possible to construct such a separating hyperplane. A very useful approach is to use non-negative slack variables

$$\eta_i \geq 0, \qquad (i = 1, 2, ..., n), \tag{13.34}$$

so that

$$y_i(\boldsymbol{w} \cdot \boldsymbol{x}_i + \boldsymbol{b}) \geq 1 - \eta_i, \qquad (i = 1, 2, ..., n). \tag{13.35}$$

Now the optimization for the support vector machine becomes

$$\text{minimize } \Psi = \frac{1}{2}||\boldsymbol{w}||^2 + \lambda \sum_{i=1}^{n} \eta_i, \tag{13.36}$$

$$\text{subject to } y_i(\boldsymbol{w} \cdot \boldsymbol{x}_i + \boldsymbol{b}) \geq 1 - \eta_i, \tag{13.37}$$

$$\eta_i \geq 0, \qquad (i = 1, 2, ..., n), \tag{13.38}$$

where $\lambda > 0$ is a parameter to be chosen appropriately. Here, the term $\sum_{i=1}^{n} \eta_i$ is essentially a measure of the upper bound of the number of misclassifications on the training data.

By using Lagrange multipliers $\alpha_i \geq 0$, we can rewrite the above constrained optimization into an unconstrained version, and we have

$$L = \frac{1}{2}||\boldsymbol{w}||^2 + \lambda \sum_{i=1}^{n} \eta_i - \sum_{i=1}^{n} \alpha_i [y_i(\boldsymbol{w} \cdot \boldsymbol{x}_i + \boldsymbol{b}) - (1 - \eta_i)]. \tag{13.39}$$

From this, we can write the Karush-Kuhn-Tucker conditions

$$\frac{\partial L}{\partial \boldsymbol{w}} = \boldsymbol{w} - \sum_{i=1}^{n} \alpha_i y_i \boldsymbol{x}_i = 0, \tag{13.40}$$

$$\frac{\partial L}{\partial \boldsymbol{b}} = -\sum_{i=1}^{n} \alpha_i y_i = 0, \tag{13.41}$$

$$y_i(\boldsymbol{w} \cdot \boldsymbol{x}_i + \boldsymbol{b}) - (1 - \eta_i) \geq 0, \tag{13.42}$$

$$\alpha_i [y_i(\boldsymbol{w} \cdot \boldsymbol{x}_i + \boldsymbol{b}) - (1 - \eta_i)] = 0, \qquad (i = 1, 2, ..., n), \tag{13.43}$$

$$\alpha_i \geq 0, \qquad \eta_i \geq 0, \qquad (i = 1, 2, ..., n). \tag{13.44}$$

From the first KKT condition, we get

$$\boldsymbol{w} = \sum_{i=1}^{n} y_i \alpha_i \boldsymbol{x}_i. \tag{13.45}$$

It is worth pointing out here that only the nonzero coefficients α_i contribute to the overall solution. This comes from the KKT condition (13.43), which implies that when $\alpha_i \neq 0$, the inequality (13.37) must be satisfied exactly, while $\alpha_0 = 0$ means the inequality is automatically met. In this latter case, $\eta_i = 0$. Therefore, only the corresponding training data $(\boldsymbol{x}_i, y_i)$ with $\alpha_i > 0$ can contribute to the solution, and thus such $\boldsymbol{x}_i$ form the support vectors (hence, the name support vector machine). All the other data with $\alpha_i = 0$ become irrelevant.

V. Vapnik and B. Schölkopf et al. have shown that the solution for α_i can be found by solving the following quadratic programming

$$\text{maximize} \sum_{i=1}^{n} \alpha_i - \frac{1}{2} \sum_{i,j=1}^{n} \alpha_i \alpha_j y_i y_j (\boldsymbol{x}_i \cdot \boldsymbol{x}_j), \tag{13.46}$$

subject to

$$\sum_{i=1}^{n} \alpha_i y_i = 0, \qquad 0 \leq \alpha_i \leq \lambda, \qquad (i = 1, 2, ..., n). \tag{13.47}$$

From the coefficients α_i, we can write the final classification or decision function as

$$f(\boldsymbol{x}) = \text{sgn}\Big[\sum_{i=1}^{n} \alpha_i y_i (\boldsymbol{x} \cdot \boldsymbol{x}_i) + \boldsymbol{b}\Big]. \tag{13.48}$$

where sgn is the classic sign function.

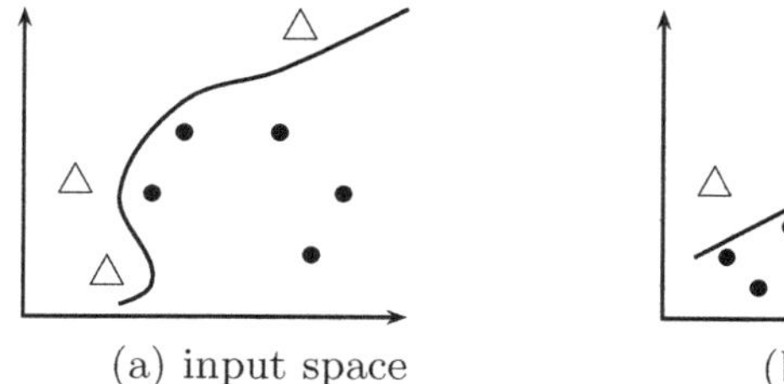

Figure 13.5: Kernel functions and nonlinear transformation.

13.2.4 Kernel Functions and Nonlinear SVM

In reality, most problems are nonlinear, and the above linear SVM cannot be used. Ideally, we should find some nonlinear transformation ϕ so that the data can be mapped onto a high-dimensional space where the classification becomes linear (see Fig. 13.5). The transformation should be chosen in a certain way so that their dot product leads to a kernel-style function $K(\boldsymbol{x}, \boldsymbol{x}_i) = \phi(\boldsymbol{x}) \cdot \phi(\boldsymbol{x}_i)$, which enables us to write our decision function as

$$f(\boldsymbol{x}) = \mathrm{sgn}\Big[\sum_{i=1}^{n} \alpha_i y_i K(\boldsymbol{x}, \boldsymbol{x}_i) + \boldsymbol{b}\Big]. \tag{13.49}$$

From the theory of eigenfunctions, we know that it is possible to expand functions in terms of eigenfunctions. In fact, we do not need to know such transformations, we can directly use kernel functions $K(\boldsymbol{x}, \boldsymbol{x}_i)$ to complete this task. This is the so-called kernel function trick. Now the main task is to chose a suitable kernel function for a given problem.

For most problems concerning a nonlinear support vector machine, we can use $K(\boldsymbol{x}, \boldsymbol{x}_i) = (\boldsymbol{x} \cdot x_i)^d$ for polynomial classifiers, $K(\boldsymbol{x}, \boldsymbol{x}_i) = \tanh[k(\boldsymbol{x} \cdot \boldsymbol{x}_i) + \Theta)]$ for neural networks. The most widely used kernel is the Gaussian radial basis function (RBF)

$$K(\boldsymbol{x}, \boldsymbol{x}_i) = \exp[-||\boldsymbol{x} - \boldsymbol{x}_i||^2/(2\sigma^2)] = \exp[-\gamma||\boldsymbol{x} - \boldsymbol{x}_i||^2], \tag{13.50}$$

for nonlinear classifiers. This kernel can easily be extended to any high dimensions. Here σ^2 is the variance and $\gamma = 1/2\sigma^2$ is a constant.

Following a similar procedure as discussed earlier for linear SVM, we can obtain the coefficients α_i by solving the following optimization problem

$$\text{maximize} \sum_{i=1}^{n} \alpha_i - \frac{1}{2}\alpha_i \alpha_j y_i y_j K(\boldsymbol{x}_i, \boldsymbol{x}_j). \tag{13.51}$$

It is worth pointing out under Mercer's conditions for kernel functions, the matrix $\boldsymbol{A} = y_i y_j K(\boldsymbol{x}_i, \boldsymbol{x}_j)$ is a symmetric positive definite matrix, which implies that the above maximization is a quadratic programming problem, and can thus be solved efficiently by many standard QP techniques.

There are many software packages (commercial or open source) which are easily available, so we will not provide any discussion of the implementation. In addition, some methods and their variants are still an area of active research. Interested readers can refer to more advanced literature.

REFERENCES

1. C. M. Bishop, *Neural Networks for Pattern Recognition*, 1995.
2. N. Christianini and J. Shawe-Taylor, *An Introduction to Support Vector Machines*, Cambridge University Press, 2000.
3. K. Gurney, *An Introduction to Neural Networks*, Routledge, London, 1997.
4. S. Haykin, *Neural Networks: A Comprehensive Foundation*, Prentice Hall, 1999.
5. V. Vapnik, *Estimation of Dependence Based on Empirical Data* (in Russian), Moscow, 1979. [English translation published by Springer-Verlag, New York, 1982]
6. V. Vapnik, *The Nature of Statistical Learning Theory*, Springer-Verlag, New York, 1995.
7. V. Vapnik, S. Golowich, A. Smola, Support vector method for function approximation, regression estimation, and signal processing, in: *Advances in Neural Information Processing System 9* (Eds. M. Mozer, M. Jordan and T. Petsche), MIT Press, Cambridge MA, 1997.
8. B. Scholkopf, K. Sung, C. Burges, F. Girosi, P. Niyogi, T. Poggio and V. Vapnik, Comparing support vector machine with Gaussian kernels to radial basis function classifiers, *IEEE Trans. Signal Processing*, **45**, 2758-2765 (1997).
9. http://www.support-vector-machine.org/
10. http://www.support-vector.net/

Chapter 14

METAHEURISTICS – A UNIFIED APPROACH

We have introduced many nature-inspired metaheuristic algorithms in this book. Some algorithms have strong similarities, while others may be directly based on or inspired by some core algorithms. Now a natural question is what possible links among these algorithms could be, and in what way? This chapter concludes this book with an intention to unify the metaheuristics.

14.1 INTENSIFICATION AND DIVERSIFICATION

The efficiency of metaheuristic algorithms can be attributed to the fact that they imitate the best features in nature, especially the selection of the fittest in biological systems which have evolved by natural selection over millions of years.

Two important characteristics of metaheuristics are: intensification and diversification (Blum and Roli 2003). Intensification intends to search locally and more intensively, while diversification makes sure the algorithm explores the search space globally (hopefully also efficiently).

Furthermore, intensification is also called exploitation, as it typically searches around the current best solutions and selects the best candidates or solutions. Similarly, diversification is also called exploration, as it strives to explore the search space more efficiently, often by large-scale randomization.

The fine balance between these two components is very important to the overall efficiency and performance of an algorithm. Too little exploration and too much exploitation could cause the system to be trapped in local optima, which makes it very difficult or even impossible to find the global optimum. On the other hand, if too much exploration but too little exploitation, it may be difficult for the system to converge and thus slows down the overall search performance. The proper balance itself is an optimization problem, and one of the main tasks of designing new algorithms is to find a certain balance concerning this optimality and/or tradeoff.

Furthermore, just exploitation and exploration are not enough. During the search, we have to use a proper mechanism or criterion to select the best solutions. The most common criterion is to use the *Survival of the Fittest*, that is to keep updating the current best found so far. In addition, certain elitism is often used, and this is to ensure the best or fittest solutions are not lost, and should be passed onto the next generations.

14.2 WAYS FOR INTENSIFICATION AND DIVERSIFICATION

There are many ways of carrying out intensification and diversification. In fact, each algorithm and its variants use different ways of achieving the balance of between exploration and exploitation.

By analyzing all the metaheuristic algorithms, we can categorically say that the way to achieve exploration or diversification is mainly by certain randomization in combination with a deterministic procedure. This ensures that the newly generated solutions distribute as diversely as possible in the feasible search space. One of simplest and yet most commonly used randomization techniques is to use

$$\boldsymbol{x}_{\text{new}} = \boldsymbol{L} + (\boldsymbol{U} - \boldsymbol{L}) * \epsilon_u, \tag{14.1}$$

where $\boldsymbol{L}$ and $\boldsymbol{U}$ are the lower bound and upper bound, respectively. ϵ_u is a uniformly distributed random variable in [0,1]. This is often used in many algorithms such as harmony search, particle swarm optimization and firefly algorithm.

It is worth pointing that the use of a uniform distribution is not the only way to achieve randomization. In fact, random walks such as Lévy flights on a global scale are more efficient. We can use the same equation (14.2) to carry out randomization, and the only difference is to use a large step size s so that the random walks can cover a larger region on the global large scale.

A more elaborate way to obtain diversification is to use mutation and crossover. Mutation makes sure new solutions are as far/different as possible, from their parents or existing solutions; while crossover limits the degree of over diversification, as new solutions are generated by swapping parts of the existing solutions.

The main way to achieve the exploitation is to generate new solutions around a promising or better solution locally and more intensively. This can be easily achieved by a local random walk

$$\boldsymbol{x}_{\text{new}} = \boldsymbol{x}_{\text{old}} + s\,\boldsymbol{w}, \tag{14.2}$$

where $\boldsymbol{w}$ is typically drawn from a Gaussian distribution with zero mean. Here s is the step size of the random walk. In general, the step size should be small enough so that only local neighbourhood is visited. If s is too large, the region visited can be too far away from the region of interest, which will increase the diversification significantly but reduce the intensification greatly. Therefore, a proper step size should be much smaller than (and be linked with) the scale of the problem. For example, the pitch adjustment in harmony search and the move in simulated annealing are a random walk.

If we want to increase the efficiency of this random walk (and thus increase the efficiency of exploration as well), we can use other forms of random walks such as Lévy flights where s is drawn from a Lévy distribution with large step sizes. In fact, any distribution with a long tail will help to increase the step size and distance of such random walks.

Even with the standard random walk, we can use a more selective or controlled walk around the current best $\boldsymbol{x}_{\text{best}}$, rather than any good solution. This is equivalent to replacing the above equation by

$$\boldsymbol{x}_{\text{new}} = \boldsymbol{x}_{\text{best}} + s\,\boldsymbol{w}. \tag{14.3}$$

Some intensification technique is not easy to decode, but may be equally effective. The crossover operator in evolutionary algorithms is a good example, as it uses the solutions/strings from parents to form offsprings or new solutions.

In many algorithms, there is no clear distinction or explicit differentiation between intensification and diversification. These two steps are often intertwined and interactive, which may, in some cases, become an advantage. Good examples of such interaction is the genetic algorithms, harmony search and bat algorithm. Readers can analyze any chosen algorithm to see how these components are implemented.

In addition, the selection of the best solutions is a crucial component for the success of an algorithm. Simple, blind exploration and exploitation may not be effective without the proper selection of the solutions of good quality. Simply choosing the best may be effective for optimization problems with a unique global optimum. Elitism and keeping the best solutions are efficient for multimodal and multi-objective problems. Elitism in genetic algorithms and selection of harmonics are good examples of the selection of the fittest.

In contrast with the selection of the best solutions, an efficient metaheuristic algorithm should have a way to discard the worse solutions so as to increase the overall quality of the populations during evolution. This is often achieved by some form of randomization and probabilistic selection criteria. For example, mutation in genetic algorithms acts a way to do this. Similarly, in the cuckoo search discussed earlier, the castaway of a nest/solution is another good example.

Another important issue is the randomness reduction. Randomization is mainly used to explore the search space diversely on the global scale, and also, to some extent, the exploitation on a local scale. As better solutions are found, and as the system converges, the degree of randomness should be reduced; otherwise, it will slow down the convergence. For example, in particle swarm optimization, randomness is automatically reduced as the particles swarm together, this is because the distance between each particle and the current global best is becoming smaller and smaller. Similarly, randomness is effectively reduced in differential evolution

$$\boldsymbol{x}_{\rm new} = \boldsymbol{x}_k + F(\boldsymbol{x}_i - \boldsymbol{x}_j), \tag{14.4}$$

whose last term is decreasing as the difference vector gets smaller and smaller.

In other algorithms, randomness is not reduced and but controlled and selected. For example, the mutation rate is usually small so as to limit the randomness, while in simulated annealing, the randomness during iterations may remain the same, but the solutions or moves are selected and acceptance probability becomes smaller.

Finally, from the implementation point of view, the actual implementation does vary, even though the pseudo code should give a good guide and should not in principle lead to ambiguity. However, in practice, the actual way of implementing the algorithm does affect the performance to some degree. Therefore, validation and testing of any algorithm implementation are important.

14.3 GENERALIZED EVOLUTIONARY WALK ALGORITHM (GEWA)

From the above discussion of all the major components and their characteristics, we realized that a good combination of local search and global search with a proper selection mechanism should produce a good metaheuristic algorithm, whatever the name it may be called.

In principle, the global search should be carried out more frequently at the initial stage of the search or iterations. Once a number of good quality solutions are found, exploration should be sparse on the global scale, but frequent enough so as to escape any local trap if necessary. On the other hand, the local search should be carried out as efficient as possible, so a good local search method should be used. The proper balance of these two is paramount.

Using these basic components, we can now design a generic, metaheuristic algorithm for optimization, we can call it the Generalized Evolutional Walk Algorithm (GEWA). Evolutionary walk is a random walk, but with a biased selection towards optimality. This is a generalized framework for global optimization.

There are three major components in this algorithm: 1) global exploration by randomization, 2) intensive local search by random walk, and 3) the selection of the best with some elitism. The pseudo code of GEWA is shown in Fig. 12.1. The random walk should be carried out around the current global best $\boldsymbol{g}_*$ so as to exploit the system information such as the current best more effectively. We have

$$\boldsymbol{x}_{t+1} = \boldsymbol{g}_* + \boldsymbol{w}, \tag{14.5}$$

and

$$\boldsymbol{w} = \varepsilon \boldsymbol{d}, \tag{14.6}$$

where ε is drawn from a Gaussian distribution or normal distribution $\mathrm{N}(0, \sigma^2)$, and $\boldsymbol{d}$ is the step length vector which should be related to the actual scales of independent variables. For simplicity, we can take $\sigma = 1$.

The randomization step can be achieved by

$$\boldsymbol{x}_{t+1} = \boldsymbol{L} + (\boldsymbol{U} - \boldsymbol{L})\epsilon_u, \tag{14.9}$$

where ϵ_u is drawn from a uniform distribution Unif[0,1]. $\boldsymbol{U}$ and $\boldsymbol{L}$ are the upper and lower bound vectors, respectively.

Typically, $\alpha \approx 0.25 \sim 0.7$. We will use $\alpha = 0.5$ in our implementation. Interested readers can try to do some parametric studies.

Again two important issues are: 1) the balance of intensification and diversification controlled by a single parameter α, and 2) the choice of the step size of the random walk. Parameter α is typically in the range of 0.25 to 0.7. The choice of the right step size is also important, as discussed in Section 4.3. The ratio of the step size to its length scale can be determined by (4.17), which is typically 0.001 to 0.01 for most applications.

Another important issue is the selection of the best and/or elitism. As we intend to discard the worst solution and replace it by generating new solution. This may implicitly weed out the least-fit solutions, while the solution with the highest fitness remains in the population. The selection of the best and elitism can be guaranteed implicitly in the evolutionary walkers.

Initialize a population of n walkers $\boldsymbol{x}_i$ $(i = 1, 2, ..., n)$;
Evaluate fitness F_i of n walkers & find the current best $\boldsymbol{g}_*$;
while (t <MaxGeneration) or (stop criterion);
  Discard the worst solution and replace it by (14.7) or (14.8);
  if (rand $< \alpha$),
    Local search: random walk around the best

$$\boldsymbol{x}_{t+1} = \boldsymbol{g}_* + \varepsilon \boldsymbol{d} \tag{14.7}$$

  else
    Global search: randomization

$$\boldsymbol{x}_{t+1} = \boldsymbol{L} + (\boldsymbol{U} - \boldsymbol{L})\epsilon \tag{14.8}$$

  end
  Evaluate new solutions and find the current best $\boldsymbol{g}_*^t$;
  $t = t + 1$;
end while
Postprocess results and visualization;

Figure 14.1: Generalized Evolutionary Walk Algorithm (GEWA).

Furthermore, the number (n) of random walkers is also important. Too few walkers are not efficient, while too many may lead to slow convergence. In general, the choice of n should follow the similar guidelines as those for all population-based algorithms. Typically, we can use $n = 15$ to 50 for most applications.

```
% GEWA (Generalized evolutionary walker algorithm)    %
% by Xin-She Yang @ Cambridge University 2007-2009    %
% Three major components in GEWA:                      %
% 1) random walk 2)randomization 3) selection/elitism %
% --------------------------------------------------- %
% Two algorithm-dependent parameters:                %
%       n=population size or number of random walkers %
%       pa=randomization probability                 %
% --------------------------------------------------- %

function [bestsol,fval]=gewa(N_iter)
% Default number of iterations
if nargin<1, N_iter=5000; end
% dimension or number variables
d=3;
% Lower and upper bounds
Lb=-2*ones(1,d);  Ub=2*ones(1,d);
```

```
% population size -- the number of walkers
n=10;

% Probability -- balance of local & global search
alpha=0.5;

% Random initial solutions
ns=init_sol(n,Lb,Ub);
% Evaluate all new solutions and find the best
fval=init_fval(ns);
[fbest,sbest,kbest]=findbest(ns,fval);

% Iterations begin
for j=1:N_iter,

    % Discard the worst and replace it later
    k=get_fmax(fval);

    if rand<alpha,
    % Local search by random walk
      ns(k,:)=rand_walk(sbest,Lb,Ub);

    else
    % Global search by randomization
      ns(k,:)=randomization(Lb,Ub);
    end

    % Evaluation and selection of the best
    fval(k)=fobj(ns(k,:));
    if fval(k)<fbest,
        fbest=fval(k);
        sbest=ns(k,:);
        kbest=k;
    end
end  % end of iterations

% Post-processing and show all the solutions
ns
%% Show the best and number of evaluations
Best_solution=sbest
Best_fmin=fbest
Number_Eval=N_iter+n

% ----- All subfunctions are placed here ---------------
% Initial solutions
function ns=init_sol(n,Lb,Ub);
for i=1:n,
ns(i,:)=Lb+rand(size(Lb)).*(Ub-Lb);
```

```
end

% Perform random walks around the best
function s=rand_walk(sbest,Lb,Ub)
 step=0.01;
 s=sbest+randn(size(sbest)).*(Ub-Lb).*step;

% Discard the worst solution and replace it later
function k=get_fmax(fval)
[fmax,k]=max(fval);

% Randomization in the whole search space
function s=randomization(Lb,Ub)
d=length(Lb);
s=Lb+(Ub-Lb).*rand(1,d);

% Evaluations of all initial solutions
function [fval]=init_fval(ns)
n=size(ns,1);
for k=1:n,
    fval(k)=fobj(ns(k,:));
end

% Find the best solution so far
function [fbest,sbest,kbest]=findbest(ns,fval)
n=size(ns,1);
fbest=fval(1);
sbest=ns(1,:); kbest=1;
 for k=2:n,
       if fval(k)<=fbest,
          fbest=fval(k);
          sbest=ns(k,:);
          kbest=k;
       end
 end

% Objective function
function z=fobj(u)
% Rosenbrock's 3D function
z=(1-u(1))^2+100*(u(2)-u(1)^2)^2+(1-u(3))^2;
% -------- end of the GEWA implementation -------
```

14.4 EAGLE STRATEGY

As we discussed earlier, a fine balance of global exploration and local exploitation is important. However, there is no strong reason why we should not use different

Eagle Strategy

Objective functions $f_1(\boldsymbol{x}), ..., f_N(\boldsymbol{x})$
Initialization and random initial guess $\boldsymbol{x}^{t=0}$
while *(stop criterion)*
Global exploration by randomization (e.g. Lévy flights)
Evaluate the objectives and find a promising solution
Intensive local search around a promising solution
  via an efficient local optimizer (e.g. hill-climbing)
  if *(a better solution is found)*
    Update the current best
  end
Update $t = t + 1$
end
Post-process the results and visualization.

Figure 14.2: Pseudo code of a two-stage eagle strategy for optimization.

algorithms for exploration and exploitation. In fact, a proper combination of different algorithms can be used for different purposes. Eagle Strategy, developed by Xin-She Yang and Suash Deb in 2010, is such an attempt, as inspired by the foraging behaviour of an eagle.

In essence, eagle strategy is a two-stage strategy. Firstly, we observe that an eagle searches for a pray by performing Lévy flights in the whole search domain. Once it finds a prey (a good solution), it changes to a chase strategy which focuses on the efficient capture of the prey (or solution). From the optimization point of view, this means that a global search is first performed by an efficient randomization technique such as Lévy flights, then an efficient local optimizer such as the steepest descent or downhill simplex methods can be used so as to find the local optimum quickly. To avoid being trapped locally, a global search is again performed, and followed by a local search. This proceeds iteratively in the same manner. This strategy can be summarized as the pseudo code shown in Fig. 14.2.

It is worth pointing that this is a methodology or strategy, not an algorithm. In fact, we can use different algorithms at different stages and at different time of the iterations. The algorithm used for the global exploration should have enough randomness so as to explore the search space diversely and effectively. This process is typically slow initially, and should speed up as the system converges (or no better solutions can be found after a certain number of iterations). On the other hand, the algorithm used for the intensive local exploitation should be an efficient local optimizer such as the Nelder-Mead downhill simplex method or the simplest hill-climbing. The idea is to reach the local optimality as quickly as possible, with the minimal number of function evaluations. This stage should be fast and efficient.

A good combination of the algorithms may be selected for a given problem, and this itself is an optimization problem. Ultimately, if an optimization system can be built in such a way that the algorithms can be selected automatically and evolve accordingly, then intelligent algorithms can be developed to solve complex optimization problems efficiently.

14.5 OTHER METAHEURISTIC ALGORITHMS

There are a few other nature-inspired algorithms that are used in the literature. Some such as Tabu search are widely used, while others are gaining momentum. For example, both the photosynthetic algorithm and the enzyme algorithm are very specialized algorithms. In this chapter, we will briefly outline the basic concepts of these algorithms without implementation. Readers who are interested in these algorithms can refer to recent research journals and conference proceedings for more details.

14.5.1 Tabu Search

Tabu search, developed by Fred Glover in 1970s, is the search strategy that uses memory and search history as a major component. As most successful search algorithms such as gradient-based methods do not use memory, it seems that at first it is difficult to see how memory will improve the search efficiency. Thus, the use of memory poses subtleties that may not be immediately visible on the surface. This is because memory could introduce too many degrees of freedom, especially for the adaptive memory use, which makes it almost impossible to use rigorous mathematical analysis to establish the convergence and efficiency of such algorithms. Therefore, even though Tabu search works so well for certain problems, it is difficult to analyze mathematically. Consequently, Tabu search remains a heuristic approach. Two important issues that are still under active research are how to use the memory effectively and how to combine with other algorithms to create more superior new-generation algorithms.

Tabu search is an intensive local search algorithm and the use of memory avoids the potential recycling of local solutions so as to increase its search efficiency. The recently tried or visited solutions are recorded and put into a Tabu list, and new solutions should avoid those in the Tabu list. Over a large number of iterations, this Tabu list could save tremendous amount of computing time and thus increase the search efficiency significantly.

Tabu search was originally developed together with the integer programming, a special class of linear programming. The Tabu list in combination with integer programming can reduce computing time by at least two orders for a given problem, comparing the solely standard integer programming. Tabu search can also be used along with many other algorithms.

14.5.2 Photosynthetic and Enzyme Algorithm

Photosynthetic algorithm was first developed by H. Murase in 2000 for finite element inverse analysis of parameter estimations in agriculture engineering. This algorithm was based on the photosynthesis in green plants. Photosynthesis uses water and carbon dioxide to produce glucose and oxygen in the presence of chloroplasts and light. The actual reaction is quite complicated, though it is often simplified as the following overall reaction

$$6CO_2 + 6H_2O \xrightarrow{\text{light}} C_6H_{12}O_6 + 6O_2. \tag{14.10}$$

In addition to light intensity and available chloroplasts, other factors controlling this reaction are temperature, concentration of CO_2, and water content. With sufficient water and CO_2, the overall reaction efficiency is largely determined by light intensity.

In the photosynthetic algorithm, the product DHAP serves as the knowledge strings of the algorithm, and optimization is reached when the quality (or the fitness) of the product no longer improves.

The enzyme algorithm (EA) was developed by Xin-She Yang at Cambridge University in 2005. It is based on the fundamental mechanism of enzyme reactions. The Michaelis-Menton quasi-steady state hypothesis of enzyme kinetics assumes the rapid, reversible formation of a complex between an enzyme and its substrate (S). The rate or velocity v is often governed by the classic Michaelis-Menton theory

$$v = \frac{V_{\max} S}{S + K_m} = \frac{V_{\max}}{1 + K_m/S}, \tag{14.11}$$

where $V_{\max}$ is the maximum velocity, and K_m is the Michaelis constant. This relationship is very similar to that used in the photosynthetic algorithm. The aim is now to optimize the product (P), similar to the DHAP in the PA. The inhibition and cooperativity, in combination with forward and backward reactions, act as an interacting buffer so as to maximize the amount of the final product. It has been applied to solve optimization problems in engineering applications.

14.5.3 Artificial Immune System and Others

There are many other algorithms we have not covered in this book. One of the most important algorithms is the so-called artificial immune system, inspired by the characteristics of the immune system of mammals to use memory and learning as a novel approach to problem solving. The idea was proposed by Farmer et al. in 1986, with important work on immune networks by Bersini and Varela in 1990. It is an adaptive system with high potential. There are many variants developed over the last two decades, including the clonal selection algorithm, negative selection algorithm, immune networks and others.

The memetic algorithm, proposed by P. Moscato in 1989, is a multi-generation, co-evolution and self-generation algorithm, and it can be considered as a hyper-heuristic algorithm, rather than metaheuristic.

Another popular method is the cross-entropy method developed by Rubinstein in 1997. It is a generalized Monte Carlo method, based on the rare event simulations. This algorithm consists of two phases: generation of random samples and update of the parameters. Here the aim is to minimize the cross entropy.

Another important algorithm is the bacterial foraging optimization, developed by K. M. Passino in around 2002, inspired by the social foraging behaviour of bacteria such as *Escherichia coli*. For details, readers can read Passino's introductory article.

Obviously, more and more metaheuristic algorithms will appear in the future. Interested readers can follow the latest literature and research journals.

14.6 FURTHER RESEARCH

14.6.1 Open Problems

Despite the success of modern metaheuristic algorithms, there are many important questions which remain unanswered. We know how these heuristic algorithms work, and we also partly understand why these algorithms work. However, it is difficult to analyze mathematically why these algorithms are so successful. In fact, these are unresolved open problems.

Apart from the mathematical analysis on simulated annealing and particle swarm optimization, convergence of all other algorithms has not been proved mathematically, at least up to now. Any mathematical analysis will thus provide important insight into these algorithms. It will also be valuable for providing new directions for further important modifications on these algorithms or even pointing out innovative ways of developing new algorithms.

In addition, it is still only partly understood why different components of heuristics and metaheuristics interact in a coherent and balanced way so that they produce efficient algorithms which converge under the given conditions. For example, why does a balanced combination of randomization and a deterministic component lead to a much more efficient algorithm (than a purely deterministic and/or a purely random algorithm)? How to measure or test if a balance is reached? How to prove that the use of memory can significantly increase the search efficiency of an algorithm? Under what conditions?

From the well-known No-Free-Lunch theorems, we know that they have been proved for single objective optimization, but they remain unproved for multiobjective optimization. If they are proved to be true (or not) for multiobjective optimization, what are the implications for algorithm development?

If you are looking for some research topics, either for yourself or for your research students, these could form important topics for further research.

14.6.2 To be Inspired or not to be Inspired

We have seen in this book that nature-inspired algorithms are always based on a particular (often most successful) mechanism of the natural world. For example, bee algorithms are based on the optimal solution of foraging and storing the maximum amount of nectar.

Nature has evolved over billions of years, she has found almost perfect solutions to every problem she has met. Almost all the not-so-good solutions have been discarded via natural selection. The optimal solutions seem (often after a huge number of generations) to appear at the evolutionarilly stable equilibrium, even though we may not understand how the perfect solutions are reached. When we try to solve human problems, why not try to be inspired by the nature's success? The simple answer to the question 'To be inspired or not to be inspired?' is 'why not?'. If we do not have good solutions at hand, it is always a good idea to learn from nature.

Another important question is what algorithm to choose for a given problem? This depends on many factors such as the type of problem, the solution quality, available computing resource, time limit (before which a problem must be solved), balance of advantages and disadvantages of each algorithm (another optimization problem!), and the expertise of the decision-makers. When we study the algorithms, the efficiency and advantages as well as their disadvantages, to a large extent, essentially determine the type of problem they can solve and their potential applications. In general, for analytical function optimization problems, nature-inspired algorithms should not be the first choice if analytical methods work well. If the function is simple, we can use the stationary conditions (first derivatives must be zero) and extreme points (boundaries) to find the optimal solution(s). If this is not the best choice for a given problem, then calculus-based algorithms such as the steepest descent method should be tried. If these options fail, then nature-inspired algorithms can be used. On the other hand, for large-scale, nonlinear, global optimization problems, modern approaches tend to use metaheuristic algorithms (unless there is a particular algorithm already worked so well for the problem of interest).

Now the question is why almost all the examples about numerical algorithms in this book (and in other books as well) are discussed using analytical functions? This is mainly for the purpose of validating new algorithms. The standard test functions such as Rosenbrock's banana function and De Jong's functions are becoming standard tests for comparing new algorithms against established algorithms because the latter have been well validated using these test functions. This will provide a good standard for comparison.

Another important question is how to develop new algorithms. There are many ways of achieving a good formulation of new algorithms. Two good and successful ways are based on two basic ways of natural selection: explore new strategies and inherit the fittest strategies. Therefore, the first way of developing new algorithms is to design new algorithms using new discoveries. The second way is to formulate by hybrid and crossover.

Another successful strategy used by nature is the adaptation to its new environment. We can also use the same strategy to modify existing algorithms and explore their new applications. For any new optimization problems we might meet, we can try to modify existing successful algorithms, to suit for new applications, by either changing the controlling parameters or introducing new functionality. Many variants of numerical algorithms have been developed this way.

In order to develop completely new nature-inspired algorithms, we have to observe, study and learn from nature. For example, I was always fascinated by the cobwebs spun by spiders. For a given environment, how does a spider decide to spin a web with such regularity so as to maximize the chance of catching some food? Are the cobweb's location and pattern determined by the airflow? Surely, these cobwebs are not completely random? Can we design a new algorithm – the spider algorithm?

Nature provides almost unlimited ways for problem-solving. If we can observe carefully, we are surely inspired to develop more powerful and efficient new generation algorithms. Intelligence is a product of biological evolution in nature.

Ultimately some intelligent algorithms (or systems) may appear in the future, so that they can evolve and optimally adapt to solve NP-hard optimization problems efficiently and intelligently.

REFERENCES

1. Bersini H. and Varela F. J., Hints for adaptive problem solving gleaned from immune networks, *Parellel Problem Solving from Nature*, PPSW1, Dortmund, FRG, (1990).
2. Blum, C. and Roli, A., Metaheuristics in combinatorial optimization: Overview and conceptural comparision, *ACM Comput. Surv.*, **35**, 268-308 (2003).
3. Farmer J.D., Packard N. and Perelson A., The immune system, adaptation and machine learning, *Physica D*, **2**, 187-204 (1986).
4. Moscato, P. *On Evolution, Search, Optimization, Genetic Algorithms and Martial Arts: Towards Memetic Algorithms.* Caltech Concurrent Computation Program (report 826), (1989).
5. Rubinstein R.Y., Optimization of computer simulation models with rare events, *European Journal of Operations Research*, **99**, 89-112 (1997).
6. Passino K. M., Biomimicry of bacterial foraging for distributed optimization and control, *IEEE Control System Magazine*, pp. 52-67 (2002).
7. Schoen, F., 1993. A wide class of test functions for global optimization, *J. Global Optimization*, **3**, 133-137.
8. Shilane D., Martikainen J., Dudoit S., Ovaska S. J., 2008. A general framework for statistical performance comparison of evolutionary computation algorithms, *Information Sciences: an Int. Journal*, **178**, 2870-2879 (2008).
9. Yang, X. S. and Deb, S., 2009. Cuckoo search via Lévy flights, *Proceedings of World Congress on Nature & Biologically Inspired Computing* (NaBIC 2009, India), IEEE Publications, USA, pp. 210-214 (2009).
10. Yang X. S., 2009. Harmony search as a metaheuristic algorithm, in: *Music-Inspired Harmony Search: Theory and Applications* (Eds Z. W. Geem), Springer, pp.1-14.
11. Yang X. S. and Deb S., Eagle strategy using Lévy walk and firefly algorithms for stochastic optimization, in: *Nature Inspired Cooperative Strategies for Optimization* (NICSO 2010) (Eds. J. R. Gonzalez et al.), Springer, SCI **284**, 101-111 (2010).

REFERENCES

1. Adamatzky A., Teuscher C., *From Utopian to Genuine Unconventional Computers*, Luniver Press, (2006).
2. Afshar A., Haddad O. B., Marino M. A., Adams B. J., Honey-bee mating optimization (HBMO) algorithm for optimal reservoir operation, *J. Franklin Institute*, **344**, 452-462 (2007).
3. Basturk B and Karabogo D, An artificial bee colony (ABC) algorithm for numerical function optimization, in: IEEE Swarm Intelligence Symposium 2006, May 12-14, Indianapolis, IN, USA, (2006).
4. Bonabeau E., Dorigo M., Theraulaz G., *Swarm Intelligence: From Natural to Artificial Systems.* Oxford University Press, (1999)
5. Bonabeau E. and Theraulaz G., Swarm smarts, *Scientific Americans.* March, 73-79 (2000).
6. Chatterjee A. and Siarry P., Nonlinear inertia variation for dynamic adaptation in particle swarm optimization, *Comp. Oper. Research*, **33**, 859-871 (2006).
7. Chong C., Low M.Y., Sivakumar A. I., Gay K. L., A bee colony optimization algorithm to job shop scheduling, *Proc. of 2006 Winter Simulation Conference*, Eds Perrone L. F. et al, (2006).
8. Coello C. A., Use of a self-adaptive penalty approach for engineering optimization problems, Computers in Industry, **41**, 113-127 (2000).
9. Copeland B. J., *The Essential Turing*, Oxford University Press, (2004).
10. De Jong K., *Analysis of the Behaviour of a Class of Genetic Adaptive Systems*, PhD thesis, University of Michigan, Ann Arbor, (1975).
11. Deb K., An efficient constraint handling method for genetic algorithms, Comput. Methods Appl. Mech. Engrg., **186**, 311-338 (2000).

12. Deb. K., *Optimisation for Engineering Design*: Algorithms and Examples, Prentice-Hall, New Delhi, (1995).

13. Dorigo M., *Optimization, Learning and Natural Algorithms*, PhD thesis, Politecnico di Milano, Italy, (1992).

14. Dorigo M. and Stützle T., *Ant Colony Optimization*, MIT Press, Cambridge, (2004).

15. El-Beltagy M. A., Keane A. J., A comparison of various optimization algorithms on a multilevel problem, *Engin. Appl. Art. Intell.*, **12**, 639-654 (1999).

16. Engelbrecht A. P., *Fundamentals of Computational Swarm Intelligence*, Wiley, (2005).

17. Fathian M., Amiri B., Maroosi A., Application of honey-bee mating optimization algorithm on clustering, *Applied Mathematics and Computation*, **190**, 1502-1513 (2007).

18. Flake G. W., *The Computational Beauty of Nature*, MIT Press, (1998).

19. Fogel L. J., Owens A. J., and Walsh M. J., *Artificial Intelligence Through Simulated Evolution*, Wiley, (1966).

20. Fowler A. C., *Mathematical Models in the Applied Sciences*, Cambridge University Press, (1997).

21. Geem Z. W., Kim J. H., and Loganathan G. V., A new heuristic optimization algorithm: Harmony search, *Simulation*, **76**, 60-68 (2001).

22. Gill P. E., Murray W., and Wright M. H., *Practical optimization*, Academic Press Inc, (1981).

23. Glover F., Heuristics for Integer Programming Using Surrogate Constraints, *Decision Sciences*, **8**, 156-166 (1977).

24. Glover F. and Laguna M., *Tabu Search*, Kluwer Academic, (1997).

25. Goldberg D. E., *Genetic Algorithms in Search, Optimisation and Machine Learning*, Reading, Mass.: Addison Wesley (1989).

26. Haddad O. B., Afshar A., Marino M. A., Honey bees mating optimization algorithm (HBMO), in: First Int. Conf. on Modelling, Simulation & Appl. Optimization, UAE, (2005).

27. Holland J., *Adaptation in Natural and Artificial Systems*, University of Michigan Press, Ann Anbor, (1975).

28. Jaeggi D., Parks G. T., Kipouros T., Clarkson P. J., A multi-objective Tabu search algorithm for constrained optimization problem, *3rd Int. Conf. Evol. Multi-Criterion Optimization*, **3410**, 490-504 (2005).

29. Judea P., *Heuristics*, Addison-Wesley, (1984).

30. Karaboga D. and Basturk B., On the performance of artificial bee colony (ABC) algorithm, *Applied Soft Computing*,**8**, 687-697 (2008).

31. Keane A. J., Genetic algorithm optimization of multi-peak problems: studies in convergence and robustness, *Artificial Intelligence in Engineering*, **9**, 75-83 (1995).

32. Kennedy J. and Eberhart R. C.: Particle swarm optimization. *Proc. of IEEE International Conference on Neural Networks*, Piscataway, NJ. pp. 1942-1948 (1995).

33. Kennedy J., Eberhart R., Shi Y.: *Swarm intelligence*, Academic Press, (2001).

34. Kirkpatrick S., Gelatt C. D., and Vecchi M. P., Optimization by simulated annealing, *Science*, **220**, No. 4598, 671-680 (1983).

35. Koza J. R., *Genetic Programming: One the Programming of Computers by Means of Natural Selection*, MIT Press, (1992).

36. Kreyszig E., *Advanced Engineering Mathematics*, 6th Edition, Wiley & Sons, New York, (1988).

37. Kuhn H. W. and Tucker A. W., Nonlinear programming, *Proc. 2nd Berkeley Symposium*, Univ. California Press, pp. 481-492(1951).

38. Lee K. S., Geem Z. W., A new meta-heuristic algorithm for continuous engineering optimization: harmony search theory and practice, *Comput. Methods Appl. Mech. Engrg.*, **194**, 3902-3933 (2005).

39. Marco N., Lanteri S., Desideri J. A., Périaux J.: A parallel genetic algorithm for multi-objective optimization in CFD, in: *Evolutionary Algorithms in Engineering & Computer Science*, Wiley, (1999).

40. Matlab info, http://www.mathworks.com

41. Michalewicz, Z., *Genetic Algorithm + Data Structure = Evolution Progamming*, New York, Springer, (1996).

42. Mitchell, M., *An Introduction to Genetic Algorithms*, Cambridge, Mass: MIT Press, (1996).

43. Moritz R. F. and Southwick E. E., *Bees as superorganisms*, Springer, (1992).

44. Murase H., Finite element analysis using a photosynthetic algorithm, *Computers and Electronics in Agriculture*, **29**, 115-123 (2000).

45. Nakrani S. and Tovey C., On honey bees and dynamic server allocation in Internet hosting centers, *Adaptive Behaviour*, **12**, 223-240 (2004).

46. Christianini N. and Shawe-Taylor J., *An Introduction to Support Vector Machines*, Cambridge University Press, (2000).

47. Octave info, http://www.octave.org

48. Pham D. T., Ghanbarzadeh A., Koc E., Otri S., Rahim S. and Zaidi M., The bees algorithm, Technical Note, Manufacturing Engineering Center, Cardiff University, (2005).

49. Price K., Storn R. and Lampinen J., *Differential Evolution: A Practical Approach to Global Optimization*, Springer, (2005).

50. Quijano N., Gil A. E., and Passino K. M., Experimental for dynamic resource allocation, scheduling and control, *IEEE Control Systems Magazine*, **25**, 63-79 (2005).

51. Reynolds A. M. and Rhodes C. J., The Lévy flight paradigm: random search patterns and mechanisms, *Ecology*, **90**, 877-87 (2009).

52. Sawaragi Y., Nakayama H., Tanino T., *Theory of Multiobjective Optimisation*, Academic Press, (1985).

53. Schrijver A., On the history of combinatorial optimization (till 1960), in: *Handbook of Discrete Optimization* (Eds K. Aardal, G. L. Nemhauser, R. Weismantel), Elsevier, Amsterdam, p.1-68 (2005).

54. Sirisalee P., Ashby M. F., Parks G. T., and Clarkson P. J.: Multi-criteria material selection in engineering design, *Adv. Eng. Mater.*, **6**, 84-92 (2004).

55. Siegelmann H. T. and Sontag E. D., Turing computability with neural nets, *Appl. Math. Lett.*, **4**, 77-80 (1991).

56. Seeley T. D., *The Wisdom of the Hive*, Harvard University Press, (1995).

57. Seeley T. D., Camazine S., Sneyd J., Collective decision-making in honey bees: how colonies choose among nectar sources, *Behavioural Ecology and Sociobiology*, **28**, 277-290 (1991).

58. Spall J. C., *Introduction to Stochastic Search and optimization: Estimation, Simulation, and Control*, Wiley, Hoboken, NJ, (2003).

59. Storn R., On the usage of differential evolution for function optimization, Biennial Conference of the North American Fuzzy Information Processing Society (NAFIPS), pp. 519-523 (1996).

60. Storn R., web pages on differential evolution with various programming codes, http://www.icsi.berkeley.edu/~storn/code.html

61. Storn R. and Price K., Differential evolution - a simple and efficient heuristic for global optimization over continuous spaces, *Journal of Global Optimization*, **11**, 341-359 (1997).

62. Swarm intelligence, http://www.swarmintelligence.org

63. Talbi E. G., *Metaheuristics: From Design to Implementation*, Wiley, (2009).

64. Vapnik V., *The Nature of Statistical Learning Theory*, Springer, (1995).

65. Wolpert D. H. and Macready W. G., No free lunch theorems for optimization, *IEEE Trans. on Evol. Computation*, **1**, 67-82 (1997).

66. Wikipedia, http://en.wikipedia.org

67. Yang X. S., Engineering optimization via nature-inspired virtual bee algorithms, IWINAC 2005, Lecture Notes in Computer Science, **3562**, 317-323 (2005).

68. Yang X. S., Biology-derived algorithms in engineering optimization (Chapter 32), in *Handbook of Bioinspired Algorithms*, edited by Olariu S. and Zomaya A., Chapman & Hall / CRC, (2005).

69. Yang X. S., New enzyme algorithm, Tikhonov regularization and inverse parabolic analysis, in: *Advances in Computational Methods in Science and Engineering*, ICCMSE 2005, **4**, 1880-1883 (2005).

70. Yang X. S., Lees J. M., Morley C. T.: Application of virtual ant algorithms in the optimization of CFRP shear strengthened precracked structures, *Lecture Notes in Computer Sciences*, **3991**, 834-837 (2006).

71. Yang X. S., Firefly algorithms for multimodal optimization, *5th Symposium on Stochastic Algorithms, Foundations and Applications, SAGA 2009*, Eds. O. Watanabe & T. Zeugmann, LNCS, **5792**, 169-178(2009).

72. Yang X. S. and Deb S., Cuckoo search via Lévy flights, in: *Proc. of World Congress on Nature & Biologically Inspired Computing* (NaBic 2009), IEEE Publications, USA, pp. 210-214 (2009).

73. Yang X. S. and Deb S., Engineering optimization by cuckoo search, *Int. J. Math. Modelling & Num. Optimization*, **1**, 330-343 (2010).

74. Yang X. S., A new metaheuristic bat-inspired algorithm, in: *Nature Inspired Cooperative Strategies for Optimization* (NICSO 2010) (Eds. J. R. Gonzalez et al.), Springer, SCI **284**, 65-74 (2010).

75. Yang X. S. and Deb S., Eagle strategy using Lévy walk and firefly algorithms for stochastic optimization, in: *Nature Inspired Cooperative Strategies for Optimization* (NICSO 2010) (Eds. J. R. Gonzalez et al.), Springer, SCI **284**, 101-111 (2010).

INDEX

Mu Sigma Inc.
3400 Dundee Road Suite 160
Northbrook, IL 60062

CPSIA information can be obtained at www.ICGtesting.com
Printed in the USA
BVOW11s0828270814
364444BV00009B/881/P